REVOLUTIONARY CHANGES IN UNDERSTANDING MAN AND SOCIETY

THEORY AND DECISION LIBRARY

General Editors: W. Leinfellner (*Vienna*) and G. Eberlein (*Munich*)

Series A: Philosophy and Methodology of the Social Sciences

Series B: Mathematical and Statistical Methods

Series C: Game Theory, Mathematical Programming and Operations Research

Series D: System Theory, Knowledge Engineering and Problem Solving

SERIES A: PHILOSOPHY AND METHODOLOGY OF THE SOCIAL SCIENCES

VOLUME 21

Scope: This series deals with the foundations, the general methodology and the criteria, goals and purpose of the social sciences. The emphasis in the Series A will be on well-argued, thoroughly analytical rather than advanced mathematical treatments. In this context, particular attention will be paid to game and decision theory and general philosophical topics from mathematics, psychology and economics, such as game theory, voting and welfare theory, with applications to political science, sociology, law and ethics.

REVOLUTIONARY CHANGES IN UNDERSTANDING MAN AND SOCIETY

Scopes and Limits

edited by

Johann Götschl
University of Graz, Austria

SPRINGER-SCIENCE+BUSINESS MEDIA, B.V.

Library of Congress Cataloging-in-Publication Data

Revolutionary changes in understanding man and society : scopes and limits / edited by Johann Götschl.
p. cm. -- (Theory and decision library. Series A, Philosophy and methodology of the social sciences ; v. 21)
Includes indexes.
ISBN 978-94-010-4165-2 ISBN 978-94-011-0369-5 (eBook)
DOI 10.1007/978-94-011-0369-5
1. Social sciences--Philosophy. 2. Anthropology--Philosophy.
I. Götschl, Johann. II. Series.
H61.R474 1995
300--dc20 95-21339

ISBN 978-94-010-4165-2

Printed on acid-free paper

TABLE OF CONTENTS

PART III

TOWARDS A SCIENCE OF MAN AND SOCIETY: PHILOSOPHICAL AND METHODOLOGICAL FOUNDATIONS

PART IV

RATIONALITY, COMPLEXITY AND UNCERTAINTY: NEW INTERRELATIONS

PART V

AESTHETICAL AND ETHICAL PATTERNS

ACKNOWLEDGEMENTS

I would like to express my deep appreciation to Werner Leinfellner. He contributed substantially in giving this book its fundamental orientation. I also want to thank Elisabeth Leinfellner, who gave us some help and advice in preparing some texts. Wolfgang Schinagl, our research assistant, created the camera-ready type version. Günther A. Höfler read and corrected the texts. Angelika Höber and Birgit Rauch, the Institute secretaries, provided invaluable services in preparing the material for print. Rebecca Welland and Angelika Peaston did a lot of translation work. My thanks go to those named here and to the authors and publisher for all their cooperation in helping us to complete this book.

INTRODUCTION

JOHANN GÖTSCHL

Over the last decades, social philosophers, economists, sociologists, utility and game theorists, biologists, mathematicians, moral philosophers and philosophers have created totally new concepts and methods of understanding the function and role of humans in their modern societies. The years between 1953 and 1990 brought drastic changes in the scientific foundations and dynamic of today's society. A burst of entirely new, revolutionary ideas, similar to those which heralded the beginning of the twentieth century in physics, dominates the picture.

This book also discusses the ongoing refutation of old concepts in the social sciences. Some of them are: the traditional concepts of rationality, for example, based on maximization of interests, the linearity of axiomatic methods, methodological individualism, and the concept of a static society. Today the revolutionary change from a static view of our society to an evolutionary one reverberates through all social sciences and will dominate the twenty-first century. In an uncertain and risky world where cooperation and teamwork is getting more and more important, one cannot any longer call the maximization of one's own expectations of utility or interests "rational".

This book presents a variety of new models which are future building blocks of a new science of humans in their societies and environments. The articles focus on recent advances in all these new disciplines. They also show their consequences for humans in their present societies, the new philosophical foundations, the new methods and the social, technological, and scientific advances. Both the criticism and refutation of traditional concepts and the construction of new models and theories give us new insights into the complex functions and structure, into the scopes and limits of our modern mass societies.

The fourteen articles, two of them written by nobelists, then, present extensions, critiques of alternatives, and foundational aspects of these new disciplines. The book shows how a cooperation of social scientists, philosophers, mathematicians, and economists can help us to make sense of the complex and very often counterintuitive and chaotic world we live in.

J. Götschl (ed.), Revolutionary Changes in Understanding Man and Society, 7–22.

CHAPTER I: CHARACTERISTIC FEATURES IN ECONOMIC SCIENCE

M. Allais: The Economic Science of Today and Facts: A Critical Analysis of some Characteristic Features

When M. Allais received his nobel prize in 1988, his fellow nobelist P. Samuelson wrote about him: "Allais is an inexhaustible source of major discoveries in economics". One such discovery certainly is Allais' new risk theory which has already radically altered today's social sciences from linear deterministic to non linear, indeterministic ones. In the present article, Allais delineates the changing methodology of today's social sciences after the invasion of stochastic and nonlinear methods.

In a time where many philosophers, including philosophers of science and New-Age philosophers, and some adherents of chaos theory as well, are fond of drawing hasty conclusions, for example, that in today's scientific practice "Anything goes", Allais rebukes their arguments. He presents the essentials of the new social sciences' methodology, which, according to Allais, begins with empirical regularities. They serve as the basis of any knowledge and any statistical forecast. In a next step, scientific models and theories erect statistical and stochastical hypotheses about these regularities and use them to test empirically all derivations from these hypotheses. A theory is devoid of any scientific interest when neither the hypotheses nor their implications can be confronted with the empirical world. Merely logical or mathematical deductions are worthless; the Golden Rule of all socially-oriented disciplines is to submit theoretical results to experiments and data.

Randomness, uncertainty and risk make it impossible for social theories to claim absolute truth. Hence, the goal of social scientists is not falsification, as Popper proposed, but to choose from existing models the "better" one. The better model will always be the one which is better confirmed by the facts and which, for a given degree of approximation, will provide the simplest representation of the observed data.

The social and moral purpose of economic activity, which has not yet been achieved, is to satisfy the unlimited needs of all societies using only limited resources, available in the form of labor, natural resources, production, and knowledge. This has to be done in an "efficient" or ethically sound way. The potential instability of contemporary economy rests on disequilibria that nobody is capable of mastering. Allais discusses the reasons why our present economic science is unable to face up to today's societal, economic and environmental difficulties.

He believes that today a new theory of human beings in their society is hampered by the growing opposition to new ideas, instigated by the tyranny of traditional doctrines. The influence of politics on social research and on social ideas obstructs the achievement of relative social and economic stability. These factors moderate growth; they block the ongoing vast global effort of all social sciences to improve our fate on earth.

G. Rosegger: Technical Change without Humans: Innovation in the Neoclassical Economic Theory

In his article, the author challenges the inability of today's neoclassical economic theories to offer an empirically satisfactory explanation of the mounting role of technological change and technological innovations.

He rejects, then, some basic assumptions of neoclassic economics: the idealized, rational utility maximizer endowed with full information and perfect foresight; the justification of decisions solely by subjective probability distributions of individuals; and, finally, methodological individualism, since methodological individualism is unable to prove that the individual actor alone plays, or can play, a decisive role vis à vis the collective, i.e., all the other actors. Consequently, Rosegger shows that it is impossible to found economics solely on the traditional individual and utilitarian preferences and on an idealized type of rationality. He demonstrates convincingly that the neglect of technological advances and dynamics in neoclassical theories is due to the theoretical exclusion of all social and human interrelations and phenomena. Morever, the traditional concept of rationality excludes the role of history, of evolutionary dynamics, and of the empirical richness of human societal interrelations. Therefore, a new and systematic understanding of technology and of its role in coping with the basic problems of modern mass societies and their environments remains outside of contemporary economics. Nevertheless, technology is one of the most important new disciplines for a future theory of human beings in their modern societies.

Because of its cultural origin, in a future science of humans technology will need a new, practically, empirically, and ecologically oriented ethical foundation.

CHAPTER II: DISCOVERY, CREATIVITY, COGNITION AND COMPUTATION: BASIC STRUCTURES

H. A. Simon: The Theory of Scientific Discovery

In a fascinating article, Simon introduces and describes a new type of "bounded rationality", a new version of his earlier "problem- or conflict-solving intelligence". Contrary to the traditional concept of rationality, bounded rationality follows inductive rules and is restricted to specific domains. It has the advantage of including the innovative power of humans. This concept of bounded rationality opposes the received view that a computer can never be innovative. For Simon, the gaining of knowledge and the process of discovery are practical, inductive processes, and inductive processes can be innovative. But this is not what philosophers understand normally by induction. Simon's new version of induction supports his view that the processes of gaining knowledge by humans and by computers show astonishing similarities. For example, putting up or inventing new models and theories is an inductive and sometimes innovative process, usually performed by human beings, but also, according to Simon, by advanced computers. It is important to know that this type of inductive mental process is by no means context-free. Only those who possess the necessary background knowledge can use, understand, operate and extend inductive reasoning creatively. Simon's computer programs do not rely on the automatic generalization of data which have been gathered in advance. Philosophers like to explain the acquisition of knowledge as generalizing. For example, one begins with "this swan is white" and "the other one is white, too" and so on, until one arrives at the inductive generalization that all swans are white. The inductive method proposed by Simon is far more creative; it is more specific and it is bounded. If somebody observes new data, then he/she will use all the already existing mathematical methods and even computational tricks in parallel. He/She will try as many as possible, combine some of them, etc., until the optimal one is found. This process is one of innovation. Simon's well-known experiments showed that very often students and established scientists reinvented already known - and famous - results (like Planck's Law), using procedures similar to the ones used by computers.

This sounds like an attack at the traditional conception of genius or creativity, but it isn't. It states simply that everybody who is skilled, sufficiently trained, and disposes of a large enough memory can be inventive or innovative. And if this is true, then even an intelligent machine which follows Simon's rules and posesses the type of rationality he has in mind can be innovative. Thus the difference between the run-of-the-mill human and the genius and, consequently, between humans and intelligent machines withers away. Simon's arguments are a powerful defense of the intelligent and creative abilities of today's "normal" science and its methods.

M. A. Boden: Understanding Creativity

Insofar as human creativity can be understood far better with the help of new computational concepts, M. Boden's approach resembles Simon's. But it differs with respect to the methods. M. Boden proposes three explanations of creativity. (i) The psychological explanation permits a new understanding of human creativity. (ii) The technological explanation is based on the attempt to produce creativity in the computer. (iii) The pragmatic explanation is founded on computers which cooperatively support human creativity.

Boden's concept of creativity rests on the realization of possible new combinations and transformations of concepts, ideas and images. Whether the novel combinations of old ideas - often called innovations - or the completely new imaginative combinations are formed in mental nets or in connectionist neuronal nets, plays a minor role.

Thus, psychology learns from combinatorial and noncombinatorial (radically transformational) creative activities how conceptual mental spaces and different ways of transformations produce new ones. Creativity can be explained as the production of new structures and ideas and their combination and transformation by generative systems within conceptual spaces.

Creative systems, be they minds or machines, function with the help of "maps", transformation possibilities, skills, and representational redescriptions.

The crucial question "Are computers creative?" can be reformulated as "Can the computer mentally perform in a way which we would regard as creative in human beings?" Boden answers this question with yes; she discusses many supporting examples, programs and models. Computer programs which create stories, like the Aesop Program and the Minstrel system, surprise us already with their unpredictable output of stories. They continue to improve; but they have not yet reached the level of a genuinely new novel.

An important area where computers help us to understand our own creativity are genetic algorithms and evolutionary programs. Both of them use random changes of their current rules, which resemble default rules, and may create entirely new forms, species, and images.

Computers can sometimes create new forms, images, ideas and concepts, and some computers can help us ourselves to do so. As a by-product, the computational approach has created many new methods and concepts. They have helped us to explain and clarify many questions about our own creative powers, and they could solve more problems than millenia of traditional wisdom. They give the psychologist a new and startling insight into the rich subtleties of the human mind.

A. Carsetti: The Role of Simulation Models in the Cognitive Sciences

It is well established that simulation models of biological languages distinguish between levels of competence (innate ability) and performance (language use) and between deep and surface structures.

A finite static system of recursive rules cannot any longer explain how generative abilities and inventions, which belong to daily linguistic practice, work. Carsetti illustrates his critical analysis of the distinction between competence and performance with the difference between Turing machines and natural languages. Such a machine can perform computations of its own program; it is also able to simulate mechanically the behavior of other machines, including itself. Carsetti uses Nelson's version of Turing machines to soften Church's definition of recursiveness by recursive functions to a new one of recursive processes. This extension, comparable to the change of traditional rationality to the weak form of rationality, should facilitate the explanation of innovations by new cognitive simulation models which Carsetti describes in his article. They surpass first order-order logic by using nonstandard, intentional (semantic) methods.

His model describes a creative "processor", furnished with new types of recursive rules which "coagulate" both levels to one processing unit. Such an integrated unit can function like a Turing system; but the unit can also go beyond it if this is necessitated by selective pressure. The model explains innovations with the help of semi-algorithms and self-organizing processes. Such a processor has a fairly high degree of complexity, strong generative and representational power, and can operate with "autopoietic" rules, similar to default rules.

Carsetti regards such a processor as a step towards an understanding and a simulation even of biological processors. Such a processor works by coupling its deep generative basis, its representational functions, and the "neural" level to one operating unit, the "coagulum".

In brief, the processor's and the neural level's creativity function basically with the help of intelligent, evolutionary dynamics of its own. Its own evolutionary dynamic is pushed forward by selective external pressure and can cope with randomness. Its innovative evolution and the biological one are two modes of evolutionary processes. Sometimes they differ, exhibiting autonomous paths of evolution; sometimes they share a common evolutionary path.

In Carsetti's highly abstract model, generative principles can be canalized, an ability which originates in the creative, semantic deep level. Therefore, any autopoietic simulation is intrinsically semantic and an integral part of natural evolution.

The purpose of this simulation model is to make comprehensible some of the necessary conditions of the creative, linguistic activity of the human neural system, especially the epigenetic processes. (Epigenetic processes are interactions between the genes and the environment that result in anatomical, physiological, and cognitive traits of organisms.)

Surely the construction of autopoietic simulation models paves the way to a future, theoretical explanation of the process of discovery. But ultimately it will depend on our future knowledge of the encoded information of the human genome and on the future realization of simulation models.

J. R. Searle: The Mind and Computation

According to a widely held theory, the brain is a computer; the mind, including its conscious aspects and mental functions, is separated from it; the mind, then, is a formal and deductive program which is implemented in the brain as the hardware. In his article, Searle develops arguments against this computational theory of mind and the dualistic view of many philosophers of mind.

Searle starts with the assumption that brains cause minds and consciousness, but that consciousness has a higher and more complex sociocultural, "mental" level. The computational theory explains the functions of our mind purely formally or syntactically. But mental states are not merely syntactical or logical; they are mental, semantic contents which need, and are supported by, the vast semantic background of our culture. For Searle, even the present cognitive science is a merely syntactic enterprise. For him, syntax is only a necessary condition, while semantics is a necessary and sufficient condition. Since semantics transcends syntax, the brain can't be a syntactic computer and the mind, a computer program.

According to Searle, computers are electronic circuits; they don't compute at all. Only then when we use them, do they compute. Computers alone do not compute intrinsically but only in cooperation with, and relative to, human users, computer designers, programers. This argumentation leads Searle to a rejection of (logico-syntactic) cognitivism. Searle's article shows a new, convincing way out of the contradictions of the mind-body dualism in its modern form.

CHAPTER III: TOWARDS A SCIENCE OF MAN AND SOCIETY: PHILOSOPHICAL AND METHODOLOGICAL FOUNDATIONS

J. Götschl: Self-Organization: New Foundations towards a "General Theory of Reality"

Götschl's article captures the philosophical consequences of the drastic changes in today's sciences. The melting together of physical theories to the grand unification theory in the sixties was followed by the unification of the natural sciences and the social sciences under the impact of the modern theories of evolution. Götschl maintains that the analysis of the scientific and philosophical conditions of this merger leads to a new science of man and his society. The emergence of new aspects and theories of evolution shook the traditional scientific view of separately and independently existing sciences. Because of the common foundation of evolutionary theories on the principle of self-organization, Götschl calls them theories of self-organization or TSO. It is already a historical fact that theories of self-organization began, in the last fifteen years, to unite natural and social sciences. This far-reaching role of creativity (self-organization) in cosmology, physics, biology, in artificial intelligence, cognitive sciences, in all social sciences, and even in the humanities led to a breakthrough. New evolutionary fundamental structures and methods challenged the received view of all traditional sciences. These new evolutionary structures differ from those of today's sciences as today's concepts differ from those of classical natural and social sciences. If our physical and social reality is the product of an ever changing, creative evolution, then philosophers have to search for a completely new understanding of our reality; consequently, they have to search for a new ontology.

Götschl's philosophical interpretation of this change gives us details of the ongoing convergence between the new natural and the modern social sciences; finally this should lead to a science of man in his society. But according to Götschl, natural and social sciences can become identical only within the unifying framework of TSO theories. Götschl shows how to overcome most of the difficulties which prevent us from understanding the fundamentals of the new "self-organizing reality" which explains so well the processual reality of man in his society. This processual reality throws a new light on the central problem: the relation between the social sciences and the natural sciences and the possible merger of both to one unified science of man and his society.

From all possible postulates which have to be taken as fundamental for this new science of man and his society, Götschl chooses self-organization as the most important presupposition of the general evolutive process which also determines social reality. Thus, the expulsion of man by the natural sciences can be stopped. The basic features of our world, the inorganic, the organic, and the cultural ones, must be of a type that allows to explain evolution and the self-organization and creativity of man and his society. This leads to the question of a new dynamic and evolutionary ontology.

Götschl's article describes the possible foundations and principles underlying the TSO theories. He has chosen the idea of self-organization, or creativity, as the new common basis of all these dynamic theories. Beside that, TSO theories have many revolutionary structures and methods in common. Contrary to the traditional sciences, these structures and concepts explain a dynamic ontology; but at the same time they also serve the unification of the natural and social sciences. This is exactly what the author calls a process-evolutive understanding of our reality. The differences between the evolutive dynamic of nature, our societies and our cultures vanish; the way for their higher unity is being cleared. Thus, the convergence of the understanding of reality - mediated by the evolutive sciences of nature - with the new social sciences produces the new science of man and his society. This philosophical reconstruction gives us an insight and a formal outline of how such a unification works.

O. Schwemmer: Towards a Science of Man

What is the real social role of humans in today's society? This is the question which has not been sufficiently answered. Neither anthropologists nor German idealistic philosophers, neither biologists nor historians nor contemporary philosophers have found a general solution to this problem. However, there are a variety of particular answers which Schwemmer criticizes one by one. Particular solutions to this fundamental question fail, however, since humans are, without doubt, the most complicated beings in existence, not only in their present mass societies, but also in the known universe.

Schwemmer's programmatic proposals for a new science of humans in their modern societies is built on a future solidarity of man with his natural world, his culture, and a plurality of societies. Culture, as the inside of man's outer world, correlates, meshes and unites the former with the latter to a single, inseparable unity. This unity has two complementary poles: the freedom of individuation and the individual's dependency on his changing environments. Their effective interconnections and role cannot be neglected within a future science of humans and their society. Such a science will become an ecological science in a twofold sense: a natural ecology and an ecology of mind.

CHAPTER IV: RATIONALITY, COMPLEXITY AND UNCERTAINTY: NEW INTERRELATIONS

W. Leinfellner: The New Theory of Evolution - A Theory of Democratic Societies

Since the eighties, a new stochastic theory of evolution has replaced, step by step, the Darwinian, biological theory of evolution; but it has also infiltrated the traditional social, economic, political and cultural theories of evolution. The article describes how this stepwise replacement took place, and how the methods of dynamic, game theoretical matrices and nonlinear differential equations became the mathematical hard core of this new theory of evolution. From the conflict-solving models of dynamic game theory, the stochastic theory of evolution borrowed new explanations of how intergenerational societal conflicts are solved in democratic societies. They do work, especially if they are dependent on randomness, uncertainty, and risk. Thus game dynamics and conflict-solving theories provide a suitable stochastic framework for the explanation of the course of societal evolution. Other disciplines dealing with the storage and traditing of information, the protection of history, the improvement of serial societal conflict solutions by learning and the concept of lotteries contribute to the theory of societal evolutions.

After its success in theoretical and empirical biology and genetics, the stochastic theory of evolution has been applied to many new fields, such as: economic, environmental, political, socio-cultural, serial, competitive and cooperative conflicts vis à vis random events, and recently to cultural and ethical conflicts.

The article shows how the stochastic theory of evolution replaced Darwinian concepts, for example, "adaptation" by "series of optimal solutions" to explain relative stable societal forms. Individual and collective intergenerational memory storage, traditing of past solutions of conflicts, and default rules enable intergenerational learning, especially within democratic societies. Utility theory under uncertainty and risk became the cornerstone of this new theory of evolution, since risk attitude and risk regulations of expected utility could replace successfully traditional expected utility maximation.

The new nonlinear, stochastic framework of evolution consists of two different methods, the evaluative, psychologically and genetically founded matrix-method and the evolution-equations and genetic algorithms. Both explain evolutionary social dynamics far better than the traditional Darwinian theory or the theory of games against nature (against random events). Leinfellner discusses the role of probability, of chance, environmental random events, societal "mutations", of evolutionary lotteries, and especially the possibility of active interference, the "canalization" of societal evolutionary process by humans.

In evolutionary processes, the maximization of expected utility plays a minor role. Relatively stable societal states are optimal solutions of evolutionary conflicts. The psychologically or even genetically rooted risk attitude is decisive when it comes to finding an optimal solution. Risk turns out to be the hidden attractor or motor of evolutionary processes; it replaces the Darwinian "survival of the fittest or the species".

Finally, some related issues are discussed, such as memory-storage and -traditing, the role of default decision rules in democratic societies, and the ongoing controversy between traditional rationality, bounded rationality, and intelligence. An appendix presents an evolutionary model.

O. Hagen: Risk in Utility Theory, in Business and in the World of Fear and Hope

Utility theory, or value theory in general, is not only the cornerstone of decision theory, game theory and of conflict-solving models, but also of micro- and macroeconomics and all social and political theories which deal with individual and collective actions, public decisions, social choice, elections and management.

In the seventies, the American School of utility, founded by von Neumann and Morgenstern, encountered a far-going criticism by M. Allais, O. Hagen, and others (the French School of utility and risk). A massive attack on the traditional type of rationality was started, especially on the maximization of expected utility: only in fictitious, idealized societies without random events, uncertainty, and risk does it make sense to follow the traditional behavioristic axioms of the Morgenstern-von Neumann school in a deductive, rational way. Since we do not live in such a deterministic, sterile world but in a real world of hope and fear, the quest for a new foundation of preference and utility theory, economics, the social sciences, and, finally, of human rationality becomes more and more important for any theory of humans in their societies.

The contribution of O. Hagen presents an overview of the new stochastic models whose foundation on risk and uncertainty shook most neoclassical theories of societies. Their new "psychological" basis anchors in our personal risk attitude which is dependent on the psychological experience of hope and fear, of future losses. The stochastic models of Hagen and of Allais and his followers, as presented in the Theory and Decision Library and Journal since 1978, show formally and empirically why the rationality ideal of neo-Bernouillian models, the basic fundamentals of neoclassical economics, foundered.

M. Bunge: Rational Choice Theory: A Critical Look at its Foundations

Bunge's paper analyzes and criticizes the foundations of the most important school in contemporary social science, rational choice theory, from a different point of view. Rational choice theory has indeed been the secret pillar of classical and neoclassical economics. Bunge denies that this theory is clear, true, relevant, and free from ideology and prejudice. It presupposes an idealized society and world. In a fashion similar to the contributions of Allais, Hagen, Munier, and Leinfellner, but from a different methodological angle, he criticizes the fundamental assumptions of traditional economic rationality. He attacks the maximation of utility and interests directly. For Bunge, the idea that social groups are mere collections of individuals and can always be reduced to individual preferences and actions seems to be circular. In sum, he opposes the foundation of economics solely on individuals and rejects methodological individualism and the principles of neoclassical and traditional economic reasoning.

After listing different types of rationality and discussions of their pros and cons, Bunge proposes a procedural rationality which is different from traditional rationality. It is the characteristic ingredient of our scientific method; at the same time, it guarantees a normative rationality as a guide to efficient actions.

Bunge concludes with the vexing problem that people do not always act rationally - in any senses of "rational" - but neither do they always act irrationally. He suggests a new type of rationality which lies between a normative-instrumental and a conceptual rationality; this form of rationality is the missing link between both extremes.

B. Munier: Complexity, the Concept of Uncertainty and Bounded Rationality of Man

In the first part of his article, Munier begins with a detailed discussion of the concepts of complexity in today's social sciences. In complex societies, the increasing interrelations between the elements (individuals, groups, parts) make it impossible to describe society by a workable linear model. Consequently, we cannot explain it with the traditional "strong" rationality, as he calls it, especially, if additional uncertainty and risk prevail.

Therefore, according to Munier, the traditional concept of a "strong" rationality in economics, which, in most cases, rests on the maximization of a utility function, has to be abandoned. "Strong" rationality, the guiding axiom of classical and neoclassical economics, leads inevitably to Allais' paradox and to other well-known paradoxes of the traditional linear or deductive models. "Strong" rationality works solely under certainty and breaks down under uncertainty and risk.

Herbert A. Simon's model of bounded rationality rests not entirely on the maximization of utility; Munier regards it exclusively as a special case of "strong" rationality. Then he discusses other variants between the extremes of a "strong" and a "weak" rationality. Since we rather seek to avoid catastrophic risks, we change from linear probabilities to nonlinear ones; this change alters both: our "strong" to a "weak" rationality and our classical, linear, deductive model to a nonlinear one.

Many other examples show that humans decide in a nonlinear fashion. In the Karmaker paradox, in the distortion of probabilities, in the lottery effect, in the fanning-out effect, probabilities are distorted to nonclassical, nonlinear ones (see also Hagen and Leinfellner). Each individual uncertainty and risk is ultimately responsible for these effects and for the nonlinearity of the models.

In a last part, it is shown that decisions based on "weak" rationality permit us to regain our control of society's economy. Moreover, Munier's proposed managerial countermeasures suggest the use of collective welfare economics. This strategy works especially in cases where uncertainty prevails, i.e., in self-organizing or truly complex systems where even negligibly small changes of the individual choices and risks may aggregate to chaotic states for the whole system. Here managerial maximin strategies are new adaptive measures; they resemble the evolutionary channeling of economic systems.

H. Lübbe: Security. On the Reasons for the Sinking Acceptance of Risk

Lübbe begins his article with the societal role and the (psychic) reasons of mounting risk and instability in our present societies. They are responsible for the leading concomitant strife for security. There is no doubt that increasing risk in modern societies has lead to a growing desire for security and stability.

Lübbe analyzes many paradoxical societal situations, for example, situations where the risky consequences of human actions become less acceptable than those of natural causes or of technical catastrophes, or when the risk of deadly accidents in street traffic is regarded as having a lower risk than atomic accidents. In most cases and for psychological reasons, catastrophic risks with great losses but with very small probabilities of their occurrence are simply neglected (see Allais, Hagen).

For Lübbe, insecurity increases with the loss of sufficient experience. The high complexity of our modern societies actually is counterintuitive, it distinguishes earlier societies from ours. Today, we have to believe experts, specialists, and managers. But their knowledge doesn't always compensate for common sense, especially when experts disagree about the risk of cumulating conflicts. Therefore, if complexity increases over a tolerable threshold, people become worried, for example: voters become confused, don't vote at all, or decide for a plain no.

Insecurity makes people feel uncomfortable; insecurity increases proportionally with the experience of impotency and hopelessness. The helplessness and ineffectiveness of people in small countries vis à vis world politics are a sign of this insecurity.

On the one hand, the experience of insecurity is mounting with the decrease of social control. On the other hand, never before in the history of civilization have we been surrounded by so many security measures which safeguard us against loss, random events, failures, robbery, etc. Paradoxically, exactly this situation increases insecurity.

According to Lübbe, this is by no means a sign of decadence. Quite the opposite is true. A fundamental change of our inner, individual values in favor of more security and stability in today's democratic societies is the reason why we support the growth of our welfare systems. The value change also reveals our longing for self-realization to an extent which is impossible in any other society. It is this emergence of the inner societal values which stimulates our strife for more societal stability and fosters our social security and stability, in spite of the rising costs of our welfare systems.

CHAPTER V: AESTHETICAL AND ETHICAL PATTERNS

P. Feyerabend: Nature as a Work of Art

Feyerabend, the author of *Against Method*, has been singled out as "currently the worst enemy of science" by two physicists in *Nature*, a charge repeated in *Scientific American*. He begins his article with his well-known skeptical, relativistic view of any scientific method. He states bluntly that neither humans, nor societies, nor even nature (whatever that is) can be captured by theories. Theories present not our objective world or truth but are solely stories about our societies and our universe. Modern scientists are the equals of myth-tellers and court jesters, as he told J. Horgan in one of his last interviews.

Feyerabend backs his view with a comparison of Parmenides' belief in a static world with Aristotle's diametrically opposed belief in a world of change. For Feyerabend, exactly the same pattern emerges again two thousand years later, in today's unresolved controversy between the static view of relativity theory and the dynamics of quantum theory. For Feyerabend, yesterday's philosophers and today's scientists develop and adhere to theories for what are ultimately only subjective and even irrational reasons. Consequently, he cannot find any objective standards by which to establish truth. Therefrom his famous dictum "Anything goes". Science, then, does not really differ from art. Feyerabend's extreme skepticism towards science is mitigated by his open-mindedness as regards other modes of knowledge. Art's symbolic representation belongs to an equally important mode of societal knowledge; in this respect, art is not very different from science, as indicated.

Theories solely describe tendencies of a particular manifest world; they resemble, or are identical with, evanescent fashions in art. The only difference is that generations worked at these theories. Taking into account that the artist and the work of art can't be separated from nature, scientists and nature can't be separated, either. That art is nature, and nature is art holds for science, too. Vice versa, we cannot separate scientists' works, the theories, from nature. Artists' creations, their works of art, and the works of scientists, but also the achievements of politicians, social reformers, and the like, who use their experience, their cunning, their personal skills, etc. are contingent. They are all equally important for the life of a society.

J. Mittelstrass: Ethics in Science - Substance or Rhetoric?

Mittelstrass focusses his article on the key question of how much ethics we need for the natural sciences, technology, and a general science of man and his society. Mittelstrass concentrates his quest for new ethical standards on the ethics of medicine. Ethics of medicine integrates most of the natural sciences, the social sciences, and technology in a concrete manner which directly determines the future of man and his society.

Traditional philosophical ethics derives ethical standards and norms from man's reason and from understanding (knowledge). An offspring of knowledge is the traditional concept of rationality which is supposed to guide our scientific and societal life.

Mittelstrass emphasizes that this type of traditional rationality is too general; it has to be replaced by a new scientific and technological rationality based on the scientist's own responsibility vis à vis the new, gigantic tasks and consequences of science and technology for our present and future societies. In a rapidly changing society and world, the consequences of traditional scientific rationality are getting out of control.

Mittelstrass illustrates with many examples how science and technology solve their own problems on the one hand; on the other hand, they create new problems. The environmental and social problems of our world cannot be solved by reducing scientific and technological progress, but only by more science and more technology. Therefore, they both have to be put under specific and new ethical standards. These new standards will become the hard core of a new ethical discipline, intrinsically connected with the sciences themselves. Mittelstrass proposes a scientific imperative as a research commandment.

Ethics of science and of technology should accomplish what pure science itself can't. Since traditional scientific and technological rationality cannot master their own problems, ethics of medicine, which lies to a great part outside of this discipline, will have to do it. The question is not whether an ethical science is possible or not, but whether it can be anchored in a general citizen's ethics or not. "Science and technology within the bounds of practical reason", then, means that the citizens have to bear the responsibility; that is the key problem of the ethics of our time.

The present book will be of interest to scholars in all disciplines. Teachers, graduate students of economics and of the social sciences, decision theorists as well as philosophers will find the chapters useful in practical life, as well as in research.

PART I

CHARACTERISTIC FEATURES IN ECONOMIC SCIENCE

THE ECONOMIC SCIENCE OF TODAY AND FACTS: A CRITICAL ANALYSIS OF SOME CHARACTERISTIC FEATURES

MAURICE ALLAIS

The purpose of this lecture is to present very briefly a critical analysis of some characteristic features of the economic science of today, as currently admitted and taught, in its relation to facts: theory and reality, pseudo-theories, the monetary and financial institutional structure of the economy, the obstacles to progress, the conditions of progress.

I. THEORY AND REALITY

Economic science, its theories and models

The essential condition of any science is the existence of regularities which can be analyzed and forecast. This is the case in celestial mechanics. But it is also true of many economic phenomena. Indeed, their thorough analysis displays the existence of regularities which are just as striking as those found in the physical sciences. This is why economics is a science, and why this science rests on the same general principles and methods as physics.

All science is based on models, whether descriptive, explanatory, forecasting, or decision-making, and every scientific model entails three distinct stages: statement of well-specified hypotheses; deduction of all consequences of these hypotheses, and nothing but these consequences, and confrontation of these consequences with observed data. Of these three stages only the first and the third - establishing hypotheses, and confronting results with reality - are of interest to the economist. The second stage is purely logical and mathematical, that is tautological, and it is only of mathematical interest.

The model and the theory it represents must be accepted, at least temporarily, or rejected, depending on the agreement or disagreement between observed data and the model's hypotheses and implications. *When neither the hypotheses nor the implications of a theory can be confronted with the real world, that theory is devoid of any scientific interest.* Mere logical, even mathematical, deduction remains worthless in terms of the understanding of reality if it is not clearly linked to that reality.

Total submission to experimental data is the golden rule which dominates any scientific discipline. It explains the extraordinary success of western

J. Götschl (ed.), Revolutionary Changes in Understanding Man and Society, 25–37.

thought in the last three centuries. This rule is the same in economics as in physics.

Any theory whatever, if it is not verified by empirical evidence, has no scientific value and it should be rejected. This is true, for example, of contemporary theories of general economic equilibrium which are based on the hypothesis of general convexity of the fields of production, a hypothesis which is disproved by all the empirical data and leads to absurd consequences. That is also the case of the neo-Bernouillian theories of expected utility which are founded on postulates whose consequences are incompatible with observed data.

The working-out of theories and their models

Abstraction plays an essential role in the construction of theories and their models. The role of science, in fact, is to simplify and to choose: to reduce facts to significant data and to seek their fundamental dependences. A mass of facts does not constitute a science. However, if abstraction is necessary, how we go about achieving it is essentially of indifference. We can simplify reality without danger and with advantage, if, and only if, this is not likely to change the actual nature of phenomena. But, in no case, whatsoever the concern for simplification, should the essence of reality change.

The legitimacy of abstraction can only be justified *a posteriori. A priori* all abstraction can legitimately be considered inadmissible. Reducing planets to points in order to study their movements is a shocking abstraction, but it works successfully, and this very success makes it legitimate. It is thus with all abstraction. This principle is valuable in economics as in any other science.

All science is a compromise between the concern for simplification and the concern for resemblance. Great simplicity is convenient, but carries the risk that the picture which emerges does not sufficiently resemble the facts; a more exact resemblance makes the model too complex and unusable in practice. What can be said is that for any given level of approximation, the best scientific model is the one which is most convenient to use.

The claim that theory and practice are opposed is completely unjustified, because a theory is valid only insofar as it constitutes a condensed synthesis of reality. If it does not, it is purely a creation of the mind, totally artificial, and of no value from the scientific point of view. If, however, the theory is actually a condensed synthesis of reality, it is extremely useful, because it represents in succinct and easily usable form a vast amount of information of all kinds on the real phenomena which it treats.

In science the notion of "*truth*" is, in fact, relative. No theory, no model, can claim to represent "*absolute truth*", and, if there is such truth, it will certainly always remain inaccessible to us. There are only models which are more or less well-verified by observed data, and of two models, the "*best*" will always be the one, which for *a given degree of approximation* will provide the simplest representation of observed data.

Whatever its empirical verification, the most we can ever say about a theory is that *everything happens as if* its hypotheses actually correspond to the real nature of phenomena concerned.

These are the general principles of the method that long ago Henri Poincaré commented on so pertinently with respect to physical sciences, and that Vilfredo Pareto so appositely extended to the social sciences.

II. PSEUDO-THEORIES

The criterion of confronting theory with experimental data is merciless. Easy as it is, with only a pen, to work out a purely literary analysis or an abstract mathematical theory as long as no empirical application is made, it is equally difficult to elaborate an analysis that is effectively verified by observed data. This doubtless explains the propensity of so many authors to avoid numerical application, except in vague and general terms.

To test the logical coherence of a theory and to bring out its real content when it deals with magnitudes linked to each other in a somewhat complex manner, mathematics is certainly a tool without equal, indeed *irreplaceable*.

But, in examining certain contemporary theories in terms of the two requirements of scientific method - logical coherence and conformity with observed data - we find two kinds of deviation: logical inconsistency, and neglect of real phenomena.

Literary theories

The defects common to many literary theories are the continual use of non-operational concepts, vague and undefined words, whose meaning changes constantly within the analysis and varies from author to author; the absence of rigour in the analysis; the abundant use of more or less metaphysical expressions, which, having no precise meaning, can mean anything one wants, and are thus sheltered from objections; the use of expressions charged with emotional content which, while they may ensure the popularity of their authors, cannot lend themselves to rigorous reasoning.

"Mathematical charlatanry"

While many literary theories cannot be considered scientific, the same can be said of a great number of theories, being purely logical, with no real link to facts. While mathematics is a tool whose mastery is extremely precious, it is, and can only be, a tool. One cannot be a good physicist or a good economist simply because one has some ability and skill in mathematics.

For almost forty-five years contemporary economic literature has developed too often in a totally erroneous direction with the construction of completely artificial mathematical models detached from reality and now it is dominated more and more by a mathematical formalism which fundamentally represents an immense regression.

Certainly, it is no longer necessary today to justify the necessity and the utility of rigourous building of models on the basis of perfectly specified axioms. However, one must be very careful not to consider that it is enough to base a theory on a rigorous axiomatization for it to be scientifically valid. Axiomatization may be necessary, but it is secondary to the confrontation of its implications in relation to experimental data.

Reading certain papers, one can only be struck by the growing abuse of mathematical formalism. There is a tendency to forget that true progress never lies in purely formal exposition; it is to be found in the guiding ideas which underly any theory. First and foremost, these ideas must be made explicit and discussed, instead of sealing them up behind a more or less hermetical veil of symbols.

Paradoxically, from the scientific point of view, incomparably more care is brought today to the mathematical elaboration of models than to the discussion of their structure, their hypotheses, and their results from the point of view of the analysis of facts.

Contemporary literature offers us countless examples of aberrations which flow from neglect of the essential principle that a theory is valid only if it is in agreement with observed facts, and that the only source of truth is experience. Indeed a large part of contemporary theoretical literature has progressively come under the control of pure mathematicians more concerned with mathematical theorems than with analysis of the real world. A new scholastic totalitarianism has arisen based on abstract and apriorist conceptions, detached from reality; this kind of "*mathematical charlatanry*" had already been denounced by Keynes in his "*Treatise on Probability*".

It cannot be repeated too often: for the economist, as for the physicist, the essential objective is not to use mathematics for its own sake, but as a means of exploring and analyzing concrete reality, and consequently never to dissociate a theory from its applications.

Unsound econometrics

But the abusive use of mathematics is unfortunately not the only failing in contemporary literature. Indeed, too often, it has generated a crop of pseudo-theories based on the mechanical application, devoid of any real intelligence, of econometrics and statistical techniques. All these theories have the same characteristics; - the elaboration of models of linear correlation which are in reality only *pseudo-models*, with a mathematical-statistical panoply of *untamed*, totally unjustified econometrics, which seem to the naive to be scientific theories whereas they are too often just empty shells; - *the blind and brutal* application of linear correlation programs and the tests associated with them, although these *tests generally are not applicable to the cases studied*; - and the use of models, too often applied to a single country and for a short period, where the explanatory variables and the arbitrary *parameters are so numerous that the obtained fittings can have practically no real meaning.*

III. THE MONETARY AND FINANCIAL INSTITUTIONAL STRUCTURE OF THE ECONOMY

The monetary and financial framework of an economy of markets

The main purpose of economic activity is to satisfy the *virtually unlimited* needs of men using the *limited* resources available of labour, natural resources, and existing equipment, given the level of technological knowledge.

But, in order that such an objective could be attained, it is necessary that the economic activity could take place *in an appropriate institutional framework*, and it is certainly an *essential* task for economic science to correctly define the general principles of such a framework.

Unfortunately experience *shows that till now, and as regards the monetary and financial framework particularly, it has not been possible to achieve this task.*

Indeed the major problem facing Western economies is the problem, completely unresolved today of the economic fluctuations and the changes in the real value of money,which endanger economic efficiency, equity in income distribution, the security of employment and resource availability, and which in the final analysis, generates social troubles. Economic instability, underproduction, iniquity, underemployment, and the distress and misery they bring about, are the worst plagues of an economy of markets. These factors are at the origin of the best-founded and most violent criticisms of societies based on economic liberty and private property.

The potential instability of the contemporary economy

Fluctuations in the real value of money make absolutely impossible an efficient and equitable functioning of an economy of markets. As a result its ethical justification may be questioned, especially as regards decentralized decision-taking, private ownership of property, and the principle of the private appropriation of surpluses.

In reality, the major inequities afflicting our society stem from the distortions of the income distribution that is brought about by changes in the real value of money.

The whole world economy today rests on gigantic pyramids of debts, each supporting its neighbour in a fragile equilibrium. Never in the past has there been such a piling up of promises to pay. Never, doubtless, has it been so difficult to cope with it.

Whether it be on currencies or stocks, frantic speculation has turned the world into a vast casino in which gaming tables are spread all over the place. Both the game and the bidding, with millions of players involved, simply never stop. Chicago quotations are followed by those of Tokyo, Hong Kong, London, Frankfurt and Paris. Everywhere speculation is made easier by the credit mechanism which makes it possible to purchase without paying or to sell without holding.

Indeed, all the difficulties met arise because a basic fact has been widely overlooked, namely that *no decentralised system founded on an economy of markets can operate properly if the uncontrolled creation ex nihilo of new means of payment allows the necessary adjustments to be postponed, at least for a while.* This means that each time expenses or debts can be paid *with mere promises to pay*, without any effective real counterpart, direct or indirect.

The effect of the credit mechanism, as it works today, and which is based on the fractional coverage of deposits, the creation of money *ex nihilo*, and the long-term lending of funds borrowed short-term, has considerably amplified the disorders that occurred. All the major crises of the 19th and 20th centuries sprang from the excessive expansion of credit, promises to pay and their monetisation, and the speculation which this expansion made possible and facilitated.

Whether foreign exchange rates or stock prices are involved, most of the time real economic data and nominal values as determined by speculation are found to move out of step with each other. Everywhere a frantic and feverish speculation is allowed, fed by, and expanded by the credit mechanism, as it works today. Never in the past has it reached such a size. The continuous quotation, minute by minute, of stocks and currencies has but one result: to considerably amplify the destabilizing effects of speculation, and to engender market manipulation and fraud.

As in all the preceding periods, the credit mechanism exercises its destabilising effects, *but on an intensified scale, both nationally and internationally*: immoderate expansion of credit, potential instability of the banking system, growth of national and international indebtedness, unstable floating exchange rates, imbalances in foreign trade and the threat of protectionism, the growth of unrestrained speculation in foreign exchange and security markets, the world use, as a value unit, of the dollar, whose real value is extraordinarily unstable and unpredictable, measures of expediency to deal with immediate threats and potential instability of the world economy.

Up to now, a collapse has been avoided, *but it is becoming ever more difficult to deal with disequilibria that nobody is really capable of mastering*. As a matter of fact, *there is no historical record of an instance in which the exaggerated ex nihilo expansion of credit has not finally culminated in its collapse.*

The disarray of economic thought

It is certainly a great intellectual and political scandal that after the periodic recurrence of major crises for at least two centuries, our democratic societies have not yet proven capable of defining economic institutions, enabling economic fluctuations to be, if not suppressed, at least considerably reduced.

From the innumerable analyses which have been produced over the past few years, only one certainty can be drawn: *the deep disagreement, indeed the disarray*, that prevails amongst experts, whether it be with regard to the floating exchange rates system, the rise and subsequent fall of the dollar, interest rates, the United States trade, payments and budget deficits, the role of the Euromarket, Third World debt, etc ... The striking fact is the absence of a consensus not just on any sort of diagnosis, but on prospects for the near future as well.

Even more significant is the fact that no one questions the very rationale of the credit system as it operates today, i.e., *ex nihilo* money creation by the banking system and the widespread policy of financing longterm credits with short-term borrowings, all factors which are eminently destabilizing.

Indeed, nationally and internationally, there is a pressing need for a full-scale rethinking of the fundamental principles on which the present monetary and financial system is based, and, as a matter of fact, an appropriate institutional structure is relatively easy to define.

Such a structure would imply at the same time a reform of the credit mechanism as it works today and an obligatory indexation of the forward contracts.

In principle, the reform of the credit mechanism should make impossible the *ex nihilo* creation of money and short term borrowing to finance loans of a longer term, and it would permit only loans at maturity terms shorter than that corresponding to the funds borrowed. Such a reform of the credit mechanism would imply the disassociation of banking activities such as they are today and their attribution to two distinct categories of establishments: - on the one hand, the deposit banks taking care of only, *at the exclusion of all loan operations*, the cash entries and payments of their clients, the corresponding expenses being billed to these later - and on the other hand, the loan banks borrowing at given terms and loaning the funds borrowed at shorter terms. In either case the competition of establishments concerned would assure their clients the best conditions possible.

Indexation of all forward contracts would permit at the same time the correctness of all economic calculations involving the future, while assuring equitable conditions for the execution of loaning and borrowing contracts. In fact, such an indexation would permit all economic agents to use a unit of account of invariable real value for all their decisions implying arbitrage between the present and the future.

Indeed, the ruling doctrines are not at all favourable to such reforms. The grip of some contemporary schools on economic thought naturally brings to mind the hold of certain religions. More power is needed to disseminate clear, simple and coherent ideas than to promote obscure concepts that, lending themselves to every interpretation, are all things to all men whatever their opinions and feelings.

The last four decades have been dominated by a succession of dogmatic doctrines, all sustained with the same assurance, each contradicting its just as unrealistic predecessor, and all abandoned under the pressure of the facts. In place of the study of history and the thorough analysis of past errors, one has been too ready to substitute simple assertions, too often resting on pure sophisms, on unrealistic mathematical modes, and superficial analyses of the circumstances of the moment.

As a matter of fact, almost all the present difficulties do result from a total misappreciation of the monetary and financial conditions of an efficient and equitable working of an economy of markets, and from an inappropriate structure of the banking institutions and the financial markets. The present incapability of economic science, as it is currently admitted and taught, to face up to these difficulties is striking.

IV. OBSTACLES TO PROGRESS OF ECONOMIC SCIENCE

I will limit myself to examining two major obstacles to the progress of economic science: the opposition to new ideas and the tyranny of dominant doctrines, vested interests and ideologies.

Opposition to new ideas and the tyranny of dominant doctrines

First, in the development of science, that is, in building theories and their models, *creative intuition always plays the determining role*. It is thanks to intuition that, from knowledge already acquired and observed facts, the selection occurs of concepts, and the relationships between these concepts that allow the essential structure of reality to be represented, that is, the choice of hypotheses. Deductive reasoning draws out all the consequences of these hypotheses and nothing but these consequences. These consequences are then confronted with the facts. Thus, creative intuition, logical deduction, and confrontation of the consequences of hypotheses with observed data, are the three basic stages in any scientific endeavour. The history of science can be characterized by the indefinite repetition of these three stages in a process which leads towards increasingly comprehensive and well-verified models.

Indeed, only through the blossoming of new ideas suggested by creative intuition and empirical evidence can science truly progress. But all real scientific progress comes up against the tyranny of the dominant ideas generated by the "*establishments*". The more such dominant ideas are taken for granted, the more they become rooted in the psychology of men, and the more difficult it becomes to gain acceptance for a new conception, no matter how fruitful it may later turn out to be. The dominant ideas however erroneous they may be, end up, simply through continual repetition, by acquiring the quality of established truths which cannot be questioned without confronting the active ostracism of the "*establishments*".

The examples of Copernicus, Galileo, Pasteur, Wegener, and so many others, demonstrate the obstacles encountered by discoverers of genius. It is this resistance to new ideas that explains why in economics we had to wait so long to discover the major contributions of Dupuit, Walras, Edgeworth, Pareto, and many others.

The successful scholar is always the one who adds some marginal improvement to the dominant theories to which everyone is accustomed. If, however, a new theory falls outside of customary channels, it is certain to face general opposition whatever its justifications.

In science, the action of the "*establishments*" and pressure groups is often exercised insidiously, sometimes moreover for reasons entirely foreign to science, and in recent years a dangerous tendency to politisation has developed in science and scientific research, based on ideological ideas of all kinds.

For all of these reasons, today just as yesterday, it is essential to subject the "*established truths*" constantly to a critical analysis without indulgence, with Pareto's statement always in mind: "*The history of science boils down to the history of errors of competent men.*".

Vested interests and ideologies

Secondly, it has been said that Euclid's theorem would have been bitterly contested had its implications brought financial or political interests into play. This is no exaggeration. By its implications the content and development of economics are very closely interwoven with vested interests and ideologies.

At any era of history, the success of the economic doctrines was ensured, not by their intrinsic value, but by the power of the interests and the feelings which the doctrines appeared to justify. Indeed, unfortunately, economic science is too often linked to the fate of political doctrines representative of vested interests and ideologies.

To take only three examples, the theory of maximum economic efficiency and the conditions it implies, the theory of the relation between prices and money supply, the theory of unemployment and its causes; these three theories have massive political implications. It matters little that a theory be confirmed, or not, by empirical data. It will be accepted or rejected according to whether its conclusions will be judged, or not, in conformity, with the interests of the dominating groups. Consequently the criterium of the truth of a theory is not anymore its conformity with the data of observation, but its conformity with the vested interests and dominant ideologies.

The theory of unemployment is particularly significant. A thorough compared analysis of the causes of unemployment in the United States, Canada, United Kingdom, West Germany and France on the one hand, and in Switzerland on the other hand, is made almost nowhere, whereas in 1987 the average rate of unemployment was 8.7 % for the first five countries and only 0.7 % in Switzerland. The only conceivable reason for this gap is that the conclusions of such a study could be very unpopular.

Further, there can hardly be a scholar who is not influenced more or less unconsciously by his own political opinions, using the word political here in its noblest meaning. This is true of every science having social implications, from history through biology to economics.

The history of the French revolution has been written by at least twenty gifted, well-known established historians, but of differing political opinions. The mode of presentation, the choice of detail, allied to an outward semblance of objectivity, give the reader of these works totally incompatible images of the same events.

Only too often the proponents of a given theory display a boundless sense of criticism of contrary opinions, at the same time as their confidence in their own theses proves that there are no limits to naivety. One of the greatest dangers in social science is a subjectiveness which refuses deliberately to take cognizance of itself.

As a matter of fact, economic science can not aim to define what ends ought to be pursued, and it would be incapable of doing so. The question of ends must be faced by every society, but their definition is no part of economic science, or of any science, for that matter.

The ends pursued may be anything. First priority may be given to economic efficiency. Alternatively, primary concern may be for a just income distribution, however relative the concept of justice may be. The aim could be to promote the fastest possible economic growth or, on the contrary, first preference might be given to stability of income and employment.

Whether efficiency, equity, or security are at issue, it is impossible to state that some ends are intrinsically preferable to others. It is for society, that is to say, in the final analysis, for the men who compose it, to decide on the form of an acceptable compromise between the different ends which may be conceived.

The ends pursued can emerge only through the functioning of the political system. Procedures differ from time to time and from country to country, but the object is always the same, namely to implement a compromise between the aspirations of the different citizens when, as is usually the case, these aspirations are contradictory.

The economist is qualified to answer only the two following questions: Firstly, are the economic ends pursued compatible with one another? Secondly, are the means resorted to appropriate to achieve the ends pursued? The object of economic science is not to specify what choices should be made. Its role is only to provide scientific information in function of which an enlightened decision can be taken.

It is an abuse, and a dangerous abuse, to let it be understood that economic science can enable scientific decisions to be worked out. An operational research model cannot provide business leaders with ready-made decisions; the best it can do is to specify their implications, given specified assumptions. The same is true of the models prepared by economic planning agencies for governmental use.

V. THE CONDITIONS FOR THE PROGRESS OF ECONOMIC SCIENCE

What are the conditions for the progress of economic science? I shall examine only two, but two very important ones indeed: concern for synthesis, and unconditional submission to the teachings of experience.

A vast global effort of synthesis

First, the field of analysis of economic science today has been so greatly enlarged, that it is tending more and more to become specialised into different fields: the description of institutions, price theory, the theory of risk, monetary theory, the theory of international trade, the theory of development, business cycle analysis, are but a few of them. There are many others.

This specialisation is necessary, for it seems clear that it would be beyond one man's capacity to have thorough knowledge of all these different fields. Nevertheless, it is essential to maintain concern for synthesis. Progress in each individual field is conditioned by a broad overall view, and thus such a view is essential.

In any event, the economic science of economics is only part of a greater whole, the social sciences. The applications of economic science inevitably have political incidences, so that the economist must also have large views in sociology, politics and history.

It is by the way from a vast global effort of synthesis that immense progress can be realized by the social sciences today. It calls very strongly for the training of economists having a broad background in history, sociology and political science, historians who are experienced in economic analysis and sociological study; and sociologists whose education has also included economics and history.

Whatever we consider, economists, historians, sociologists, should all have a good knowledge of the mathematical and statistical techniques for the elaboration of models, of their conditions of use, and of their limits.

The unconditional submission to teaching of experience

The second condition for the progress of economic science, like any science, is the search for objectivity and the unconditional submission to teaching of experience.

If a science is to progress, the judgements and decisions of those who administer, study, teach, or apply it, must strive for objectivity. If total objectivity is certainly impossible, perhaps we can try to get as close to it as possible? The essential property of error is one's belief to hold the truth. Thus he who thinks he is right may be making a mistake. Indeed he who makes a mistake is doubly ignorant. He does not know the right answer and he does not know that he does not know it. The more reasons we have for thinking we are right, the more we ought to be convinced of the relative nature of any conviction, and the readier to accept opinions which differ from our own.

It is never possible to pronounce finally on men or things. In the sciences, the propositions which have been considered by each century as the best established have given way to others which in their turn have been replaced under the pressure of facts. This is one of those regularities which can be extrapolated with full confidence. Some opinions which the world today takes for unquestionable will be recognised tomorrow as untenable.

The major principle of scientific discipline is always to be prepared to doubt that which one considers true, to be ready to examine favourable opposing views, and to further research, designed to disprove propositions with which one agrees.

Doubt as to one's own opinions and the respect of others' are the precondition of any real progress in science. Universal, or even majority agreement, cannot be considered as criteria of truth. In the last analysis, the essential condition for the progress of science is total submission to the teachings of experience, the only real source of our knowledge.

As my master Vilfredo Pareto said: "It is always the concrete phenomenon which decides whether a theory should be accepted or rejected. There is and can be no other criterion of a theory's truth than its more or less perfect agreement with concrete phenomena.".

Prof. Dr. Maurice Allais
15, Rue des Gate/Ceps
92210 Paris, France

REFERENCES

1966 "L'Economique en tant que Science (Economics as a Science)", Lecture given at the opening session of the academic year 1967-1968 of the Institut Universitaire de Hautes Etudes Internationales, of Geneva, October 23th, 1967, *Collection des Etudes et Travaux de l'Institut Universitaire de Hautes Etudes Internationales*, no. 6, 26 p. Reproduced in the *Revue d'Economie Politique*, January 1968, pp. 5-30. English translation: "Economics as a Science", *Collection des Etudes et Travaux de l'Institut Universitaire de Hautes Etudes Internationales*, no 7, 24 p. and *Cahiers Vilfredo Pareto*, 1958, pp. 5-24.

1971 "Les Théories de l'Equilibre Economique Général et de l'Efficacité Maximale - Impasses Récentes et Nouvelles Perspectives (Theories of General Economic Equilibrium and Maximum Efficiency - Recent Blind Alleys and New Perspectives)", *Revue d'Economie Politique*, May-June 1971, no. 3, pp. 331-409.

1974 "Theories of General Economic Equilibrium and Maximum Efficiency", in: *Equilibrium and Desequilibrium in Economic Theory*, D. Reidel Publishing Company, Dordrecht, 1977, pp. 129-201. (English translation of Allais 1971. The English version only differs from the French one in a few marginal additions, particularly in note 18).

1976 *L'Impôt sur le Capital et la Réforme Monétaire*, (Tax on Capital and the Monetary Reform), Editions Hermann, Paris, 1977, 370 p., second edition 1988.

1979 "The So-called Allais' Paradox and Rational Decisions under Uncertainty", in: *Expected Utility Hypothesis and the Allais' Paradox*, Contemporary Discussions of Decisions under Uncertainty with Allais' Rejoinder, Maurice Allais and Ole Hagen editors, Reidel Publishing Company, Dordrecht 1979, 714 p., pp. 434-698.

1984 "The Credit Mechanism and its Implications", in: *Arrow and the Foundations of the Theory of Economic Policy*, George R. Feiwell, edit., Macmillan Press, 1987, 758 p., pp. 491-561.

1987 *Les Conditions-Monétaires d'une Economie de Marchés - De la Réflexion sur le Passé à la Préparation de l'Avenir*, (The Monetary Conditions of an Economy of Markets - From the Reflection on the Past to the Preparation of the Future), Montchrestien 1989.

1988 *My Life Philosophy*, The American Economist 1989.

1989 *Autoportraits* (Self-Portraits), Montchrestien, Paris 1989.

TECHNICAL CHANGE WITHOUT HUMANS: INNOVATION IN THE NEOCLASSICAL ECONOMIC THEORY

GERHARD ROSEGGER

I.

Those who are familiar with conventional economic theory, based on the neoclassical paradigm, may well ask whether this theory should be counted among the "sciences of humankind". More than three decades ago, Kenneth Boulding provided an incisive answer to this question:

> Economics has been, and still largely is, a sunless astronomy of commodities, seeking an ever changing equilibrium in cycles not quite as regular as those of the planets, but still moved by differential equations rather than by men.[1]

If this astronomy had its Kepler, it would be the nineteenth-century French economist, Leon Walràs, whose *Elements d'economie politique pure* laid the foundation for the imposing structure of neoclassical theory. This would, however, be the Kepler of the *Mysterium cosmographicum* of 1595, which was based on the assumption of circular planetary paths. For while Kepler was willing later to to revise his theory to accommodate the empirical observations of Tycho Brahe, which clearly pointed to elliptical paths, the neoclassical paradigm has largely remained immune to all evidence contradicting its assumptions and conclusions.

Heterodox approaches have managed, at best, to cause some fissures in the neoclassical structure, without being able to shake its foundations. In observing this fact one has to admit, however, that none of these approaches in its present stage of development has been able to counterpose a logically closed system that matches the orthodoxy - thus leaving itself open to the accusation of "ad-hocism".

Acute epistemological difficulties arise in several specific areas. One of these is the neoclassical theory's inability to offer an empirically satisfactory explanation of the origins and effects of technological change. Its treatment demonstrates clearly that a seemingly unbridgeable gap exists between the standard textbook version of the phenomenon and observed economic (and political) reality. In his entirely sympathetic review of the neoclassical paradigm's approach to technology, an author concludes:

> The outcome of all this is to argue that there are strong theoretical reasons to believe that from economic analysis we can make predictions as to the determinants of the rate of technological advance and the nature of technological advance in different industries, as well as saying something

J. Götschl (ed.), Revolutionary Changes in Understanding Man and Society, 39–51.

about the bias of technological advance at the macro-level. However, empirical support for many of these theories is not really as encouraging as one might hope.[2]

The following observations are intended to point out some reasons for this state of affairs, which is hardly a cause for self-complacency on the "queen of the social science's" part.

II.

One could base a critique of the neoclassical paradigm on a variety of epistemologically equivalent considerations. For heuristic reasons it is most convenient to focus on the axioms underlying all (equilibrium) models derived from the paradigm:

1) All actors in the economic game are "rational" utility maximizers, endowed with full information and perfect foresight. (In another contribution to this book, Professor Bunge discusses the dubiousness of this assumption).
2) Where such perfection is lacking, actors make their decisions on the basis of (subjective) probability distributions; these can be described by decision-theoretic models based on the laws of statistics. (Bypassing abstract debate about this axiom, March and Shapira [3] have shown that the observed behavior of decision-makers bears no resemblance to the theoretical postulate).
3) Each actor is so insignificant relative to the collectivity of all actors that his decisions provoke no identifiable reaction. (The assumption that each actor plays a "game against nature" may be tenable under some conditions; but these are precisely the conditions under which it would be irrational to seek advantage by means of differentiated behavior, e.g., through innovations).

Given the above suppositions, self-organizing market forces are taken to produce an equilibrium with globally maximized utility - a situation in which no actor can improve his utility without impairing that of another actor. In other words, this is a static equilibrium in the sense of Newton, a condition which can be "disturbed" only through exogenous factors. The sources of innovations are assumed to lie outside the market mechanism and they are therefore taken to belong to these exogenous factors. As one critic observed, this treatment of technological advance turns the whole system into a performance of "Hamlet" without the Prince of Denmark.

Because of its very structure, neoclassical theory has nothing to say about the social origins of new technical knowledge or about how this knowledge is diffused throughout an economic system. But even within the framework it has defined for itself *a priori*, the theory cannot ignore the effects of the "disequilibrating disturbances" set in motion by the innovative behavior of actors. The dilemma arising from the need formally to recognize the *impact* of technological change *upon* the economic system provides the basis for a more specific critique.

Besides the postulate of rational, utility-maximizing "decision units," the construct of *the production function* forms a centerpiece of the neoclassical explanatory construct, in micro- and macro-economics. The function describes output, for the economy as a whole as well as for individual organizations (enterprises, plants), as the result of the application of only two inputs - capital and labor. Therefore, the actual processes of production take place inside "black boxes," whose changing technical and institutional (organizational) characteristics can be described via the observation of shifts in (relative) input and output quantities.

In macroeconomic studies, technological progress appears as an unexplained residual in the aggregate production function, i.e., as an increment in an economy's total real output that cannot be explained through increases of the capital or labor input. Although this approach raises serious methodological questions,[4] econometric investigations (and comparisons) of countries' economic growth continue to be based on the assumption that such models can describe the effects of technological progress. The exercise could be regarded as harmless, were it not for the fact that government decision-makers frequently use the results as indicators of the need for various policy measures.

Setting aside this issue, Fritz Machlup has also argued convincingly[5] that the explanation of macroeconomic phenomena cannot be regarded as satisfactory unless it is capable of demonstrating what types of microeconomic processes generated the observed, aggregate behavior of a system. Neoclassical theory fails this test, because at the level of the firm or plant it can do no more than to register shifts in the relative shares of capital and labor and in the input-output coefficients (factor productivities) ... under the assumption of fully informed, optimizing decision units.

Friedlaender used an example to demonstrate the inability of this approach to make meaningful statements.[6] She asks how econometricians, drawing on the neoclassical production function, would explain the effects of two recent innovations in the automobile industry - the introduction of so-called "quality circles" and of "just-in-time" delivery of components. The first innovation presumably raises the marginal product of labor relative to that of capital, while the second reduces working-capital requirements. Depending on the relative weight of these effects on the flow of production, econometricians could describe the (combined) innovations as labor-saving, neutral, or capital-saving. Friedlaender argues that such descriptions tell nothing about the nature of these technological changes or about the ways in which they influenced the organization of production.

Attempts to add to this paradigmatically dictated apparatus any deeper hypotheses about the origins and implementation of innovations are bound to fail, from a methodological perspective, because they would require abandonment of the optimization axiom. The formalization of this axiom requires a careful separation of those variables which a decision-maker includes in the calculation of the benefits and costs of a decision from those that are not included in the calculation, but which nevertheless have more or less significant impact on society. Environmental effects immediately come to mind in this connection, but these can be fitted into the neoclassical paradigm's frame of reference via the theory of "market failure". From the social-science viewpoint, more significant are the difficulties of conventional theory in *explaining* all the, unintended, *network effects* that are triggered by innovations - instead of simply registering them *ex post facto*. A theory of "socio-technical systems" adequate to this task is only in the early stages of development.

Given its logical foundations, there is no reason why the rational decision unit of neoclassicism would ever venture onto the shaky ground of radical innovation. In an exploration of this topic, the authors render a precise description of the dilemma: "One appreciates the non-rational nature of the innovation process when one notes that the more novel the invention is, the less orderly and predictable is the process.".[7] Thus, for example, even John Maynard Keynes - in this respect firmly rooted in the neoclassical tradition - could do no better than to explain investment behavior that was, according to standard decision models, irrational by means of a scientifically not very satisfactory recourse to the "animal spirits" of decision-makers.[8]

III.

It would be a mistake, of course, to condemn neoclassical theory as irrelevant for economic reality because of its failure to offer an adequate explanation for the origins and significance of technological change. The theory's most important contribution no doubt lies in the careful circumscription of the necessary and sufficient conditions for the static (allocative) efficiency of an economic system and its units. These conditions are not only the foundation for the structure of modern microeconomics but also for many operational and tactical decision models of management science. Even here, however, indiscriminate reference to the rationality assumption has produced some epistemologically questionable results, as, for example, in the current practice of referring to enterprise strategies as "optimal".

One must recognize also that neoclassical doctrine created a vocabulary and a grammar without which any contemporary discourse would have to be conducted in a much impoverished language. Nevertheless, this too raises a bothersome question: to what extent does the adoption of a standard vocabulary and grammar affect the modes of thought and scientific strategies even of heterodox theorizing? [9]

Like all successful scientific programs, the formal theory of allocative efficiency rests on a procedural foundation whose elements can be suggested here only in caricatured brevity:

- Fixed pre-suppositions - e.g., income distributions that are given *a priori,* as are resource endowments, technologies, and the preference functions of consumers. (How the latter acquire such functions for innovative products, about whose characteristics they are not yet informed, must remain unexplored).
- Positive heuristic rules - e.g., the relentless quest for optimization in all economic activities and the interpretation of participants' behavior in an optimization framework. In scientific praxis, the application of this rule leads to the development of logically consistent, tightly-woven mathematical structures, which form a protective wall against any attacks that cannot be framed in the same context.
- Negative heuristic rules - e.g., avoidance of historical, institutional, and humanistic arguments, as well as the consignment of questions of ethics and social justice to the "extraeconomic sphere".
- Methodological rules - e.g., interpretation of functional relations among economic units that act according to the rules of rational behavior (or of rational expectations) in terms of a tendency toward equilibrium. (The postulate of decisions made by "units" leaves open the question about what kinds of intra-organizational processes - obviously involving the individual, subjective evaluations of members - lead to the "objective rationality" of these decisions. This would exist only if one assumes that all members of the organization have identical preference functions ... or that the organization is run by an autocrat).

The theory's normative elements derive from these procedures. They also explain why it is necessary to assign attempts at explaining the motivation for technological advances to the realm of "dynamic efficiency," about which the neoclassical theory can only make statements that are far from the static situation's logical consistency.[10] Special problems are created by the phenomenon of cooperation in technical development by enterprises that otherwise compete vigorously in the marketplace. Not too surprisingly, the question whether such cooperation is efficient (welfare-enhancing) or inefficient (rivalry-reducing) leads to seemingly endless debate but cannot be answered definitively within the neoclassical context.

The static conception of competition, in which abstract and anonymous "market forces" produce a social optimum, inevitably leads to an elevation of atomistic (perfect) competition to the role of yardstick for measuring the efficiency of real-life markets. In his *Theory of Economic Development* (1912) Schumpeter attempted to show that technological progress is not produced by (statically efficient) markets but that it has to be attributed to the activities of the type he called "innovator-entrepreneur," who acts against the equilibrating tendencies of existing markets.

In view of the heuristic rules outlined above, one can understand why the neoclassical mainstream managed to divert this attack by pointing out that Schumpeter's innovator-entrepreneur played a role in the drama of economic development that was - despite its sociological interest - scientifically inadmissible, because it was the role of a *deus ex machina*. Only much later (1941) did Schumpeter undertake a massive reformulation of the dynamically relevant concept of competition, in the process putting the neoclassical, static view of efficiency through market rivalry in its proper place:

> The competition that counts is the competition from the new commodity, the new technology, the new source of supply, the new type of organization ... competition which strikes not at the margin of the profits and the outputs of existing firms, but at their foundations and their very lives. This kind of competition is as much more effective than the other [neoclassical] as a bombardment is in comparison with forcing a door, and so much more important that it becomes a matter of comparative indifference whether competition in the ordinary sense functions more or less promptly... .[11]

Although this revolutionary formulation of market dynamics is recognized today by most textbooks, at least in a footnote, it has not really influenced conventional models and their evaluation.

IV.

As suggested above, the logical consistency of the neoclassical system - derived from its presuppositions - goes a long way toward explaining its dominant position. Although they can point to a host of empirical and historical observations that contradict the assumption of a tendency to general equilibrium inherent in real-life economies, heterodox schools of thought have, so far, not been able to displace the conventional paradigm.

The collective strength of these schools lies in their efforts to build existing knowledge about the complexity of *human* motivation and *human* behavior into explanations of the economic process. Thus, for example, theory of "bounded rationality" and of "satisfying", prominently connected with the work of Herbert Simon, puts into question the postulate of optimizing economic units. As mentioned above, even the introduction of formal decision-theoretic models, based on the rules of statistics, is incapable of rescuing this postulate. Especially in the case of strategic decisions, including decision to innovate, the statistical distributions on which these models are based exist only in textbooks, not, however, in the heads of those who are actually confronted with such decisions.[12]

Similarly, organization theory has pointed out the untenability of the conception of an enterprise as a servomechanism whose inputs and outputs are connected only by a transfer function with shifting coefficients. And the "new institutionalism" has shown to what extent the way in which a society organizes its economic system influences the system's dynamic performance. The conclusion here is that, except in their most rudimentary form, the "laws" of economics are not invariant with institutional arrangements and incentives.

In this connection, approaches resting on the paradigm of evolution (and ecology) have yielded some very insightful results. Kenneth Boulding must be mentioned as a pioneer in organizing the approach.[13] Nelson and Winter[14] have undertaken what must so far be considered the most comprehensive attempt to draw on the evolutionary paradigm in constructions of a micro- and meso-economic theory of economic change.

And yet, despite these attacks as well as forays by economic historians, the neoclassical paradigm continues to guide mainstream theorizing as well as teaching and research at most universities. The reason for this state of affairs has to be sought in the fact that neoclassicism still offers the only *logically closed*, completely formalized system of thought. By now it is so routinized as to be easy to teach but nevertheless sufficiently challenging to test students' mettle; this, too, may help to explain its resilience!

Now, the closing of a system of thought is always necessary if it is to make some "generally valid" statements. But this general validity then applies only to the territory within the system's boundaries, as defined *a priori*. Logical consistency alone does not require that the mapping of the territory be a projection of reality, reduced to its true key elements. In this sense, economic theories are always bound to be incomplete insofar as they are forced to abstract from the multiplicity and complexity of social dynamics. One is forced to ask, however, whether technological change, as a key element in these dynamics, should not be shown even in a very abstract map of the economic landscape.

In his critique of the neoclassical school, Henry Woo[15] proposes a simple test: is a closed system capable of enriching its fundamental theses through additional, compatible hypotheses? And he argues that especially conventional economic theory cannot do this, because its dense mathematical formalization does not permit the absorption of non-conforming hypotheses, especially hypotheses derived from the empirical observation of human behavior. Therefore, Woo finds the theory "closed but incomplete" in a sense that is epistemologically unacceptable.

V.

If one relates this criticism to the obvious impossibility of opening the neoclassical system to the phenomena of technological change, one is forced to conclude that the criteria of allocative efficiency (i.e., of optimization under known input constraints) are bound to fail, because in a given state of the economy it is technology that defines what is considered an input. Hence it is necessary to distinguish two fundamentally different types of innovation:

a) Induced innovations, which represent an entrepreneurial reaction to shifting (relative) input prices and which are therefore motivated by market signals[16] These can be fitted into the neoclassical production-function framework, albeit with a bit of methodological stretching.
b) Innovations of a higher order (basic innovations), which revolutionize the concept of resources and therefore do not just alter the relevant coefficients of the production function but bring entirely new parameters into the game. This is especially obvious in the case of natural resources[17], but it applies with equal force to the definition of the human input into production (know-what and know-how). It is precisely this type of innovation, which does not have to be technically brilliant but which nevertheless evidences the genius of the innovator (a human being, not a "decision unit"!), that gives the impetus for society's long-run dynamic.

Attempts to hack the Gordian knot of historical change, which cannot be captured in rigid models, by means of time-invariant "capital units" and "labor units" are bound to fail. To define what is left of progress in terms of a qualitatively vague, residual in the aggregate production function is no more than a convenient methodological fiction. *De facto* this kind of reductionism turns technical progress, one of the main driving forces of history, into no more than an unexplained component of an equation purporting to describe economic change.

An opening of the neoclassical system to a recognition of this historical fact surely would undermine its logical consistency (and its mathematical structures). The system simply has no room for truly epochal technical breakthroughs, those that lead to the destruction or foundation of whole new sectors and with them to profound social transformations. These effects are subsumed under the concept of continuous "structural change". A misinterpretation of the motto of Alfred Marshall, one of the fathers of neoclassicism, - "*Natura non facit saltum*" - may well have contributed to this view of the history of technology. Therefore, representatives of the neoclassical school also deny categorically the possibility that the clustering of basic innovations in time could produce "long waves" of the kind first hypothesized by Kondratieff.

Beyond all of this, an opening of the system would also require recognition of the fact that the motivation for innovative behavior derives not only from the striving for economic benefits or for an increase in economic efficiency. From a historical and broader social-science perspective, technology develops in a field demarcated by the poles, culture, politics, and economics; and furthermore, extra-social, non-cognitive human behavior no doubt also plays a role in pushing out the frontiers of "knowledge applied in production". Consequently, the notion of a competitively motivated *laisser innover* must be extended in (at least) two directions.

First, it has to be recognized that even individually motivated innovations are not just the result of a quest for efficiency, however that may be defined, but that they are also an expression of a spirit that is best described by the concept of *homo ludens*. In retrospect, we recognize that "playing" with technical ideas, without any economic incentives or effects, nevertheless enriches the inventory of societies' total stock of knowledge about technology[18].

Second, the opening of the system in the direction of culture and politics requires an acceptance not only of individualistic but also of societal motives for innovation, as they are implemented by politicial decision-makers (in the broadest sense of that term). It would be wasteful to enlarge, in this connection, on the significance of state-induced innovations in armaments. But in the form of so-called "spin-offs" even these affect other areas of technology, including those dominated by a pure market orientation. Extreme advocates of the vision of a "market" for all new scientific and technical knowledge[19] abstract from this uncomfortable consideration. The market analogy is certainly stressed beyond the breaking point in all those cases in which the state plays the role of initiator and financier of technological development as well as that of sole customer (monopsonist) for its products.

These observations refer not only to the obvious case of military technology but also to a broader range of innovations stimulated by a variety of organizations. The quest for national prestige, ritualism, sports, and even charitable efforts should be mentioned in this connection, although these topics cannot be pursued in detail. It seems clear, in any event, that a theory of self-organization must not restrict itself to the market mechanism if it wants satisfactorily to explain the origin and effects of innovations.

Moving from the positive to the normative realm, some special difficulties arise from the neoclassical paradigm's claim to exclusiveness. This has to do, once again, with the necessity of treating technological change as a phenomenon exogenous to the basic model of the economic process - despite all contrary efforts and despite bows in the direction of Schumpeter and other heterodox views. Such a conception almost inevitably leads to a *reification* of technology, as though it played its role independent of the social drama.

Thus, the fundamentally cultural origin of technological advance is obscured[20] and it becomes tempting to interpret technology as the source of all intellectual and societal ills[21] or as the panacea for all these ills. On this basis, it then appears a manageable task to draw on a more or less authoritarian form of "technology-push" for the achievement of goals that are in the interest of society as a whole.[22] The assumption derives from a recognition of the fact that "socially responsible" behavior frequently is difficult to motivate; this insight then leads to a conclusion hardly justified by empirical evidence: that it is therefore more efficient to obviate the motives and effects of asocial behavior by technological means.

Here is a challenge that transcends the boundaries of purely scientific critique. If the neoclassical paradigm is accepted as a leitmotiv for political action, it is bound to produce disappointments to the extent that such action is based on the exogenous, and therefore directly manipulable, character of technology. The prevailing school of economic thought cannot meet the demands of providing normative guidance:

> In the United States, the Congressional hearings on science and technology policy, and the speeches of high officials on these matters, are full of 'facts' discovered by economists about the major role of technological advance ... While all the attention recently given by politicians to scholars is flattering, we believe that the scholarly community has much less to say about appropriate policy toward innovation than many scholars like to believe. Prevailing theory of innovation has neither the breadth nor the strength to provide much guidance regarding the variables that are plausible to change, or to predict with much confidence the effect of significant changes.[23]

From this observation, one may venture the pessimistic conclusion that the nexus between technology and the collective behavior of humans probably cannot be captured adequately by economic science - even if a paradigm shift were to occur. The complex network of causation and effects requires a system-oriented perspective, which would *a priori* challenge the claim to exclusiveness of economics in explaining and steering technological change. Only by assuming that technological change is "pre-programmed" according to its own laws, could one speak of an "adaptation" of social events to the conditions posed by science and technology. And only under the equally questionable assumption that causation runs entirely in the opposite direction, could one yield to the illusion that technological advance may be organized and directed through centralized decisions. The dubiousness of both extreme positions, which either interpret all relationships among humans *in vacuo...* or which project all relevant relationships onto a plane with the coordinates of the human material world has been pointed out, for example, by Götschl. [24]

VI.

It was the purpose of these arguments to show that the treatment - or rather, the non-treatment - of technological advance in neoclassical theory can be explained by an epistemological fact: the theory owes its analytical pliability and its logical tightness to the deliberate exclusion of all social phenomena that fail to fit into its structure. The price it has to pay for this appearance of "scientific" solidity is its inability to include in its models the historical and empirical richness of human and societal experience. Hence, a systematic understanding of technology, in its role of forming societies and at the same time being transformed by them, per force remains outside the theory's scope.

Whether this closure of the orthodox system of thought stems from a desire to establish dubious analogies with the natural sciences, especially with (Newtonian) physics, or whether it can be explained by science-sociological factors along the lines of Kuhn's thought, must remain an open question. It is an established fact, however, that even Alfred Marshall, generally regarded as the most "scientific" precursor of the modern neoclassical system, pleaded for a biological (i.e., dynamic and evolutionary) economic science, though the main body of his own work rested mainly on mechanistic-static foundations.[25] His efforts to coordinate rigorous models with a realistic description of the key phenomena of a developing economy were at best partially successful. On German and Austrian academic soil, the same intellectual conflict was played out, albeit in a slightly different form, in the famous-notorious "Methodenstreit".

If one moves beyond the epistemological and methodological problems of neoclassicism, one is forced to conclude that the roots of its difficulties are essentially philosophical. They lie in a purely individualistic and utilitarian conception of human behavior in the economic sphere. Without this conception, and without the axiom of "rationality" derived from it, every optimization model, no matter how mathematically complex, would remain a chimera.

No normative conclusions should be derived here from this observation. It is meant only to substantiate the observation that there is no room in the neoclassical intellectual structure for human striving after adventure - and therefore for innovation - if one takes this structure at face value and does not weaken it through scientifically unacceptable, adaptations to the real social world.

When Stanley L. Jaki writes about "the three faces of technology",[26] he points to the impossibility of fitting the phenomenon into the framework of a single discipline, without a general philosophical perspective. At the same time, he criticizes the kind of pseudo-humanism that therefore regards all theories about technology as nothing but more or less successful "opinions" which - like all fashions - are bound to come and go. In his view, the issues we confront in connection with radical technological advances must be resolved by philosophy, and especially by ethics. And he warns scientists and technologists not to gloat over the disagreements among philosopers. For, so argues Jaki, whatever else may engage our minds in the way of theoretical themes about technological change, there remains a fundamental question for humankind. In the face of technology's rapidly increasing complexity and the resulting threat of a tyranny of expertise, free societies must answer this question not only in scientific but in very practical ethical terms. It is: *Quis custodiet ipsos custodes*?

The question tends to encourage modesty about our understanding of social dynamics, not only among economists of the neoclassical or other persuasions but also among all others who seriously seek to come to terms with the essence of technology as the most *human* manifestation of the quest for knowledge.

Prof. Dr. Gerhard Rosegger
Weatherhead School of Management
Case Western Reserve University
Cleveland, Ohio, USA

NOTES

1 Boulding, K.E., "Ecology and Economy", in: Sanford, G.W. (ed.), *Segments of the Economy: A Symposium*, Cleveland, 1957, pp. 9-25.
2 Stoneman, P., The Economic Analysis of Technological Change, Oxford, 1983, pp. 62.
3 March, J. G. and Shapira, Z., "Managerial perspectives on risk and risk taking", *Management Science, 33/11* (November 1987), pp. 1404-1418.
4 This topic is elaborated in Baird, R.N., "Production functions, productivity, and technological change", in: Gold, B. (ed.), *Research, Technological Change, and Economic Analysis*, Lexington, MA 1977; pp. 11-41.
5 Machlup, F., *Der Wettstreit zwischen Mikro- und Makro-Theorie in der Nationalökonomie*, Tübingen, 1960.
6 Friedlaender, A.F., "Macroeconomics and microeconomies in innovation: The role of the technological environment", in: Landau, R. and Rosenberg, N. (eds.), *The Positive Sum Strategy*, Washington, DC 1986.
7 Price, W. and Bass, L.W., "Scientific research and the innovative process", *Science, 164/3881* (May 16, 1969), pp. 802-806.
8 Keynes, J.M., *The General Theory of Employment, Interest, and Money*, New York, 1936; pp. 161.
9 On the topic, "methods as languages", see the contributions of Boulding, Knight, and Samuelson, with a commentary by F. Machlup, in: "Issues in methodology", *American Economic Review, XLII/2* (May 1952), pp. 34-73.
10 Cf. Machlup, F., "Statics and dynamics: kaleidoscopic words", *Southern Economic Journal 26* (October 1959), pp. 91-110.
11 Schumpeter, J.A., *Capitalism, Socialism, and Democracy*, New York, 1942, pp. 84-85.
12 March and Shapira, op. cit.
13 Boulding, K.E., *Evolutionary Economics*, Beverly Hills, CA 1981.

14 Nelson, R.R. and Winter, S.G., *An Evolutionary Theory of Economic Change*, Cambridge, MA 1982.

15 Woo, K.H., *What's Wrong with Formalization in Economics: An Epistemological Critique*, Newark, CA 1986.

16 On this topic, see Binswanger, H.P and Ruttan, V.W. (eds.), *Induced Innovation: Technology, Institutions, and Development*, Baltimore, MD 1978.

17 This argument is fully elaborated in DeGregori, T. R., *A Theory of Technology: Continuity and Change in Human Development*, Ames, IO 1985.

18 Examples and interpretations can be found in Daedalus of New Scientist (anon.), "Pure technology," *Technology Review* (June 1970), pp. 38-45.

19 Rottenberg, S., "The economic approach to science policy", *International Journal on the Unity of the Sciences, 1/4* (Winter 1988), pp. 387-398.

20 Gordon, T.J., "The feedback between technology and values", in: Baier, K. and Rescher, N. (eds.), *Values and the Future*, New York 1969.

21 Despite all contemporary versions of fundamental critiques of technology, the most all-embracing is still Juenger, G.F., *The Failure of Technology*, Hinsdale, IL 1949.

22 See, for example, Weinberg, A.M., "Can technology replace social engineering?", *University of Chicago Magazine 59* (October 1966), pp. 6-10.

23 Nelson, R.R. and Winter, S.G., "In search of a useful theory of innovation", *Research Policy 6* (1977), pp. 36-76.

24 Götschl, J., "Wissenschaftlicher Fortschritt und Bedingungen für Humanitätsgewinn (Aspekte eines Humanismus der wissenschaftlich-technischen Kultur)", *Wissenschaftsforschung, 4/2* (September 1988), pp. 9-23.

25 Marshall, A., *Principles of Economics*, 8th edn., New York 1920.

26 Jaki, S.L., "The three faces of technology: idol, nemesis, marvel", *Intercollegiate Review, 23/2* (Spring 1988), pp. 37-46.

PART II

DISCOVERY, CREATIVITY, COGNITION AND COMPUTATION: BASIC STRUCTURES

THE THEORY OF SCIENTIFIC DISCOVERY[1]

HERBERT A. SIMON

Professor Leinfellner has already sketched out the general strategy of Artificial Intelligence. I have just a few words to say about it. It's a field that is about thirty-five years old which has always moved along two lines: one having to do with finding ways of getting computers to do intelligent tasks, to do tasks which, if they were done by human beings, would call for our human intelligence. And that's the direction of research we usually call "Artificial Intelligence". There is a closely related, a closely allied direction of research which consists of trying to understand how the human mind does intelligent things; how we human beings solve problems; how we use language; how we make decisions; how we design objects; create new objects of one sort or of another. And this direction of research is often called today "Cognitive Science". In carrying it out and trying to answer those questions, we have found the computer a very powerful assistant, a very powerful tool because today characteristically we try to state our theories of how we human beings think in the form of computer programs. The computer program is our mathematical language for stating formal theories of how human beings think. And so there has been a very close relation over these thirty or thirty-five years between Artificial Intelligence, which seeks to empower computers to do things that had required human intelligence, and Cognitive Science, which consists of using computers to model that human intelligence. In my remarks tonight, I'm going to be directed primarily in a second direction in asking what we have learned about human thinking by this attempt to model it with the computer. Now, research in this field adopts a very obvious program, one that we adopt in all areas of science. First, we see whether we can explain simple things; and as we begin to gain some confidence that we are understanding those simple things, we move on to things that are more complex or that appear more complex.

And so early research in this field was concerned with puzzle-like tasks, tasks that require intelligence of human beings because we find that intelligent human beings may take a good deal of time to do them, but puzzles of the sort that can be used in the psychological laboratory. We can bring subjects in; we can explain a puzzle to them; we can ask them to try to solve the puzzle; we can ask them to think aloud while they are solving it.

We think today that we understand pretty well the sequences of information processes that go on in the human head when the human being is solving a puzzle-like task. I don't know whether you have the puzzle in Germany, whether you all learn it as children, the problem of the missionaries and cannibals:

J. Götschl (ed.), Revolutionary Changes in Understanding Man and Society, 55–73.

There are three missionaries, three cannibals on one side of the river with a boat. You want to get them on the other side of the river. But there is a problem: the boat will only hold two people, and if you get more missionaries than cannibals in any place, the missionaries will eat the cannibals, or perhaps it's the other way around: the cannibals will eat the missionaries. Those are the kinds of problems which we studied in intelligent human beings who hear that problem for the first time. It may take 15 minutes or more to solve it. It is obvious that we can write computer programs to solve that problem; but we think we can write computer programs which solve it in the manner using the sequences of steps that humans do. And notice, I've said: "Use the sequence of steps or use the information processes that humans do." We still know very little about how neurons accomplish this task. We still know very little about the biological aspects of intelligent performance and the linkage between our psychophysiological knowledge and our knowledge of the information processes - it's still a very weak linkage. But we should not despair from that. This is a state of affairs we see in all of the sciences. Chemists operated for a very long time in developing chemical theory; they carried chemical theory very far during the 19th century and well into the present century without really having any help or very much help from the physicist as to the underlying structures of atoms, the internal structure of atoms, and the elementary particles which, at several layers lower down, we know today explain the chemical phenomena.

Science can be built in layers in which we explain different phenomena at different layers of complexity, and, fortunately for us, science can sometimes be built from the top down, hanging, you might say, from sky-hooks. So we could have a successful chemistry to a very important extent before it had a foundation in physics; and today we can learn a great deal about the information processes going on in the human head without knowing how those processes are accomplished by neurons, without knowing what electrical and chemical processes support them. So it's that upper level of information processes I'll be talking about.

I have briefly indicated the strategy of the research going from these puzzles to tasks which can only be solved by people who have a considerable amount of knowledge about the subject matter. For example, there has been a great deal of research on the game of chess - and no one plays championship chess without a tremendous amount of knowledge about chess. We think today that we have a fairly complete theory of the information processes that people use in a domain like chess; and that encourages us to move on to another area of problem solving which we call "ill-structured problem solving".

There are many problems where the goals of solution are only incompletely stated, where, when we start to attack the problem, we only have vague indications of what means are available for attacking it, of what path we might follow in attacking it. An obvious domain of ill-structured problems is the problem of doing architectural design or engineering design where the full scope of the problem only emerges in the process of solving it. We start out with a very incomplete statement of our goals and gradually elaborate those goals as we go along.

Another area that is very unstructured is the area of scientific work, or, as I will talk about it tonight, scientific discovery. At least we, as scientists, like to think that the work of scientific discovery is not cut-and-dried, that the problems we initially tackle are not sharply defined problems with sharply defined methods of solution, at least in the most important and most creative science; and so the study of scientific discovery is the study of one form of ill-structured problem solving, open-ended, if you like, problem solving. Now, we approach this research with a hypothesis. The hypothesis I have already indicated is that discovery is a form of problem solving in that the processes which we have discovered and learned about in the study of simpler forms of human problem solving will turn out to be the processes that are needed when human beings do more creative things, when they do the kind of things that are involved in scientific discovery.

Now, that's an empirical hypothesis; that's nothing that we have to decide here and now. We don't have to accept, and we don't have to reject. It's an empirical question whether we can provide a satisfactory explanation of the processes of scientific discovery, using only the basic processes that we used in explaining simpler forms of human thinking and human problem solving. Now, what are those basic processes, those basic processes for manipulating information or for manipulating symbols? They're, in fact, very simple processes. We think that what's required is the ability to take in, or to input, information from external sources - let's call it "reading"; the ability to take information from inside and send it outside, transmit it outside as I'm trying to do now - which we'll call "writing" or could be called "speaking". We think we need a capability for storing not only individual symbols or patterns but for storing complicated sets of relations among symbols or patterns in some kind of a memory. We think that it needs to be a capability of manipulating and modifying those structures that sit in memory. We think there is needed a capability of comparing patterns like the pattern of my fingers, comparing them to see whether the two patterns are the same or different and then behaving one way if they are the same and another way if they are different.

So the hypothesis, more specifically stated, is that a system which is capable of performing these elementary functions has the basic resources it needs for intelligence and, in fact, the basic resources it needs for the kind of intelligence that solves scientific problems and makes scientific discoveries. Now, we know that computers have these properties. We know computers can read symbols, write symbols, store symbol structures, modify symbol structures, and compare structures. We know computers can do that. The hypothesis is, first, that, since computers can do that, they are capable of intelligent behavior and of modeling the intelligent behavior of human beings, including such behavior as making scientific discoveries; and second, that, because we human beings have the same capabilities, we're able to do those things.

And let me emphasize again that I put this forward as an empirical hypothesis; it's to be tested now, it's to be tested, first of all, by looking at the data we can get on human scientific discovery. A lot of it and a lot of the data I will refer to is historical data. But we also can - and I will give some examples - discover, to a certain extent study, scientific discovery in the laboratory. So, on the one hand, we are going to try to gather together data about human capabilities of making discoveries. On the other hand, we are going to write computer programs which will model those processes of making discoveries. And, of course, if they model them successfully, those programs will also be able to make discoveries; or, at least, at the very least, given the same initial conditions as a human scientist, the same data that the human scientist starts with, the same problem that the human scientist starts with, if our theories are successful, then our programs should be able to take those initial conditions and recreate the discovery of the human scientist. And it's chiefly evidence of that kind that I'm going to be talking about in the remainder of my remarks tonight. We'll see as we go along and as those of us who are scientists already know that there are many different kinds of science; I don't mean just many different fields of science but many kinds of scientific problems and many approaches to scientific problems. At first I'd like to talk about what is usually called "data-driven science", and I have already indicated roughly what it is. Sometimes, not always, a scientist will be confronted with a body of data - maybe the data that he got himself or herself, maybe the data that were published in a journal or presented by some kind friend. But we start with some data, and then the question is: are the data lawful, can we find regularities in that data, can we sum up the data in a neat scientific law, perhaps a mathematical function or some other kind of law?

And I'd like to talk a little bit about data-driven science and to talk specifically about a computer program which we called "BACON", named after Sir Francis Bacon, who, of course, had some theories of his own as to how scientific laws could be induced from data. The work I want to describe is joint work over almost a decade with my colleagues Pat Langley, Gary Bradshaw, and Jan Zytkow.

Let me start with some work done by a local hero here in Graz by the name of Kepler. I think the actual discovery that I want to mention was made a little after he left Graz; maybe he was already thinking of it. Well, he was thinking of part of it, we know, while he was here.

Kepler, as you know, discovered many important things, one of them what we now call "Kepler's Third Law". He had available to him data, pretty good data, on the distances of the planets from the sun and on the periods of revolution of the planets around the sun. And he observed - and I think others had also - that the farther a planet was from the sun, the longer it took to revolve around the sun. Now, the question was: could one find a more precise law to describe that relationship? And Kepler, after a long period, did in fact describe the law that we call today "Kepler's Third Law". In the first part of his work on that problem, he, in fact, came to the conclusion that the period varied as a square of the distance from the sun. Today we don't accept that law, and if you look at his data - well, he got a pretty good fit. But clearly there were discrepancies between the data and his law. It's always a good question as to when you are satisfied with how well your data fit the theory, and your theory fits the data. For a while, he was satisfied. He was busy with other things. As you know, since you live in the land of Kepler, his mother during this period was being tried for witchcraft and he necessarily spent a certain amount of his time worrying about that and trying to do something about it.

But after a while, he became dissatisfied with his solution to the problem. He decided the fit wasn't good enough, and he went back to it; and in a short period of a few weeks he found the law we now call "Kepler's Third Law".

Now, we have given that same problem to the computer program BACON. In fact, we have given it the actual data that Kepler had when he started out on the problem. BACON, I should observe, does not proceed by trying all possible hypotheses. In the first place, it will be very hard to define what you mean by "all possible hypotheses"; in the second place, by any reasonable definition, there are quite a few - like several - infinities of possible functions that you could try to apply to data. So, clearly, if BACON has to have any chance of success, it must do something other than simply trying things at random - and it does. It follows several simple rules. But these rules have nothing to do with the subject matter of the data that it's treating. That's why we call this "data-driven science". Kepler had essentially no clues from physics as to what that function ought to be. The relevant physics, the relevant explanation in terms of gravitation didn't come along until Newton.

So, what Kepler started with were these raw data. He had a few theological notions, too; but I'll leave these out at the moment. What BACON does is to notice that the periods get longer as the distances get larger. And BACON has a rule of thumb, or heuristic, as we call it, which says: when two quantities vary together - both getting larger together - then try dividing the one by the other. You might be lucky: The ratio might be a constant, and you really found a law. Of course, it usually won't be. But you can try. It's very cheap, in a computer at least, to divide or to compare numbers by each other, but a little harder for Kepler: who did not have logarithms at the time he started working on this problem, or even when he finished it.

In this case, of course, if you divide the period by the distance, you don't get a constant. But you can keep those numbers around, and then you can do the same process over again, using the new numbers you have as well as the old. You might now notice that the ratio of period to distance also grows with the distance and divide again. What you then get is the ratio of the period to the distance square. But you'll ask: is it also not constant if you take a strict rule of constancy, that is, if you demand a rather good fit to the data? Notice that the second function which BACON tries is the law, or assumed law, which Kepler first discovered. It's not the hundredth function or the thousandth function; it's the second function which BACON tried, following a very simple-minded rule. If we set the standards of equality in BACON tight enough so that it's not satisfied with that function, BACON's program has to make that decision. If we set it tight enough, then BACON will simply go on. It will now discover that the ratio of p to d squared gets larger as the ratio of period to distance gets smaller. And now, instead of dividing the one by the other, it multiplies the one by the other; and the result is Kepler's Third Law. So by using some extremely simple rules for searching, but searching with extreme selectivity, not at random - it's a great selectivity - the program quickly finds Kepler's Third Law. I can say: well, good luck, fine, we found a law. The question is whether the same tricks will work if you put it to the test of other laws. And so we have tried BACON out on a great many scientific discoveries of the 18th and 19th century; and some it will not do, and others it will do.

For example, consider the following experiment. We take two heavy objects and fasten them to a spring and stretch it. We release the spring and measure the accelerations. We do this again and again, stretching springs sometimes very wide, sometimes just a little bit. BACON, taking those data on the accelerations, will quickly discover again that the ratio of the acceleration of the two bodies is always the same. Now take two new bodies. Do the same. BACON finds the same law but, of course, with the different ratio. The ratio depends on which two bodies you take. Now, BACON has a third rule of thumb which it applies in instances where it finds a law in the form of a relation between two bodies, or two situations, more generally. What BACON will do in that circumstance is to postulate that the reason the law holds is that each of these bodies has some property, and that the relation is due to the interactions of properties of the two bodies. And so BACON begins to assign numbers - I won't go through the arithmetic of it - to the two accelerated bodies and now finds that it can explain all the results with as many bodies as you please, with the constant numbers assigned to each of the bodies. It doesn't have to postulate a different acceleration ratio for each new pair of bodies because it can calculate that acceleration ratio by simply knowing this number for each of the bodies. Now, the physicists in the audience - maybe those who aren't physicists - will have already guessed what these numbers are, that is, what we human beings call the number. But BACON has invented or reinvented - it's known to us - the concept of inertial mass, a very central concept, of course, in classical physics.

Now just notice what went on here - I don't want to have sleight of hand. The data that BACON was given were only data on accelerations; it was simply given measures of the accelerations of these bodies. It was BACON's heuristics, its rules of thumb, which lead it to hypothesize. And it might be able to state the law more simply if it assigned the proper properties to the two bodies and led it to calculate what those properties should be; and it was in the course of doing that, that BACON introduced the new concept of inertial mass. In the same way, given data on mixtures of hot liquids, the initial temperatures, and equilibrium temperatures of mixtures of hot liquids, BACON discovers the concept of specific heat. In other experiments - I should keep saying "rediscoveries" - it discovers the concept of index of refraction of light as light goes from one medium to another and is bent as a result. So here we have a demonstration at least of a process where we'll - given nothing but some raw data - attempt to find, and often succeed in finding, lawful relations among the data. And we'll in many cases, in order to state those relations more parsimoniously, more elegantly, invent new concepts, which turn out in these instances to be very central and important concepts in physics.

Now, I haven't said anything about human performance, and let me turn to that for a moment. Well, I have said something about Kepler. But I had not provided you with any data of human performance. Let me take a different example; and it happens to be an example which BACON in its present form will not do, although the changes in BACON required to do it are fairly trivial. A gymnasium teacher of geometry in a Swiss high school in - I think - the year 1883 went to a lecture being given by a physics friend of his. The physics friend talked about new data that had been discovered on the wave length of the hydrogen lines in the spectrum of light from the sun. He mentioned four lines that had been measured, four wave lengths that had been measured; and he observed in his lecture that these wave lengths appeared to have some sort of lawful relation. They were spaced out in such a way that, if you just looked at them, you'd have a feeling it might be lawful. But nobody knew what the law was. There's, of course, no theory whatever in physics which would explain why there should be any lawful relation among the lines of the hydrogen spectrum from the sun. Balmer was the person who is interested in playing with numbers and with figures. He took these numbers, copied them down, and took them home; and in about two months he found a lawful relation among them, again a rather simple one. If you arrange the lines in order - the first, second, third, fourth, n-th line - then it turns out that the wavelength of the n-th line is simply the square of n divided by the square of n minus four; and Balmer arrived at that result. We know a little bit about how he did it. Some of his papers have been preserved, and there have been a couple of historical studies made on them. The papers really don't reveal much detail of the process.

Now one question you can ask about this is at least the question of whether Balmer had to do something extraordinary in order to arrive at this result. We can't answer whether he did something extraordinary because we don't know exactly what he did. But there was something extraordinary required to come to this result. Or would anybody who is sufficiently motivated, at least any bright person who is sufficiently motivated, to work on this problem for a while find Balmer's results? So we put this to experimental test.

We hired a graduate student (in the United States as elsewhere, graduate students are usually always very glad to have some additional income) for the summer term, a little time; and we told him that we had some numbers which we'd like him to examine and see whether he could find any regularity in those numbers, whether he could find a law governing those numbers. And we said that we were also a little interested in how he go about it. So, would he, please, carry a tape recorder around, and would he, please, carry a notebook around, and whenever he worked on this problem, would he write down his work in that notebook? And he agreed to these terms of employment and went to work. Oh yes, I think, we also said to him that, although we were hiring him full time, we really didn't expect him to work more than twenty or thirty hours a week on this problem (we were afraid, he might go crazy if he did), that he could use the rest of his time for his thesis research, or to go to the beach, or for anything he wanted to do. He went to work on the problem, and from time to time we met with him just to see how things were going; and he soon became very frustrated. He found himself making no progress at all on the problem. But he was getting a good wage, and he was a good sport, so he kept working at it. At first, some weeks after he had started, I don't think he managed to keep up even the thirty hours a week; it was too frustrating. After he had put in about fifty-five hours of work on the problem of manipulating these numbers, playing with them, trying to fit functions to them, trying to divide them by each other and altering them, and so forth and so on, he performed a manipulation which made the numbers look a little simpler than they'd been before. He was very encouraged by that, he worked without stopping for five hours, and at the end of those remaining five hours, he had Balmer's law.

Now, we don't know that he did it in exactly the way that Balmer did. But we do know, as with others who experiment, that it does not require anything extraordinary to do it because we know in considerable detail exactly what he did. We know what functions he examined, what order he examined them in, and the nature of the simplification that lead to his solution; and I can, without going into details, simply say that there was nothing extraordinary about the things he did: he did just what all of you would do if you are confronted with that problem. He looked at the numbers; he graphed them; he saw they dropped off; tried an exponential; tried a lot of algorithmic curves; tried polynomials; and for a long while didn't get anywhere; played with the numbers themselves: could he normalize them by dividing all of them by the first number? - and so on.

They're just the things we would do, and sixty hours of work did it. Now, there are some interesting unanswered questions about Balmer's motivation. Since no one was paying him to do that, he must have been highly motivated to play with numbers - and he did. Let me take another human example where we have some both historical and laboratory evidence. Oh, by the way, I should mention that, of course, the Balmer discovery was another example of pure data-driven discovery. As to the theoretical reasons why that particular function fits the data; we are only elucidated by Bohr in 1913, thirty years after Balmer had done his work. In fact, in Bohr's first published paper on the Quantum Theory of the atom, it was Balmer's formula which provided one of the main pieces of empirical evidence for Bohr's result. So, as in the Kepler-Newton case, so in the Balmer-Bohr case data came first, and empirically derived, induced theory came second; and a deeper, theoretical explanation of why that function fit came third and very much later - not always true in science but true there. Well, let's take one more example.

In the middle of the 19th century, a few physicists began to be interested in the so-called black-body problem - and again, if you don't already know, never mind what the black-body problem is; it has to do with radiation in certain cavities which are called "black bodies". And the interesting question, interesting to physicists, is: as you change the temperature of that cavity, what happens to the wave length of the radiation in the cavity? Do the wave lengths change in that cavity? By the middle of the 19th century, a little bit was known about the problem, but nobody knew what the correct function was. At the same time, experimenters were working and were very greatly improving the instruments for measuring black-body radiation and extending the range of wave lengths over which they could measure them. By the nineties, they had some very good data over a considerable range of wave lengths, and a law was announced, again empirically fitting functions to these data, which came to be known as "Wien's law" (after a distinguished German physicist) in 1896, which essentially said that the function was an exponential - never mind an exponential of what, but you all know what exponential curves are: they start out slow and rapidly accelerate. The law was an exponential.

A young German physicist, Max Planck, was just starting his career at that time. He knew a great deal about thermodynamics, he knew a great deal about electromagnetics. He had even begun to study the work of Boltzmann, the foundations of statistical mechanics, which were completely new and unknown at that time. He decided that the way to start his career was to provide a theoretical explanation for why Wien's law held, his law that was just known to hold because it fit the data. And so he began to build a model, a theoretical model of a black body in terms of the known thermodynamics and the electromagnetism in the cavity, and applying a little bit of Boltzmannian statistics; and at the beginning of 1900, at the turn of this century, he was satisfied that he had proved from basic classical physical principles that Wien's law was Wien's law because it had to be that the mechanisms of a black body simply determine the radiation in such a way that it had to fit Wien's law. And he proudly - I think it was proudly - published a paper to that effect in a leading German physical journal in January of 1900.

But the experiment is weird work, and as theorists know, experimenters always cause trouble. You get a nice theory, and then an experimenter comes along and says: but there is this fact over here, and that doesn't seem to fit your theory. And the experimenters here were continuing to extend the range of black-body radiation into the infra-red and now could make new measurements. And, alas, these new measurements did not fit Wien's law at all; they did not come even close. On October 7th of 1900, Planck's friend Rubens came with his wife to call on the Plancks - a Sunday afternoon call. And, of course, very soon the two physicists were in a corner talking about their common interests in physics - the world has not changed in too many respects in these years; and Rubens announced to Planck that he now had data which made Planck's Wien's law and Planck's explanation of it completely unacceptable. In fact, the radiation near the origin, far from being an exponential, seemed to be linear through the origin, whereas for larger values it certainly looked a kind of exponential. Before he went to bed that night, Planck had written down a new law that fit the data through its whole range, and, in fact, mailed a postcard to Rubens to prove that he'd done it. And that law is the law of black-body radiation that we accept down to the present day, and it's the foundation stone of quantum mechanics, of quantum theory at first and then quantum mechanics.

But notice that Planck was still in a state of great embarrassment because the preceding January he had proved that the other law had to be right in terms of physical principles. And so now Plank went to work and examined his proof in great detail to see if he could find a weak spot and correct it. Of course, he knew the result he had to come out with; he knew that result the first time through, too. And he found a step in his theory that had to do with resonators, with feeding this little system of resonating bodies where it seemed to him he could tinker with it a little bit. And so he applied what he thought was a new Boltzmannian statistical mechanics. He had to do a couple of funny things to apply it, knowing what the result was going to be. He had to, in fact, make the problem discrete and introduce a little constant into the calculations which he did not have any other reason for doing; and he came out with a right answer. He came out with the function which is now the law of black-body radiation; it's an exponential with a little number minus one tucked on to the end of it. And there we were. The explication of what this little constant was he introduced came much later. Really, it was Einstein and Ehrenfest in 1905 who first called major attention to it and pointed out that it was not simply a computational quirk, but that it was at the very center of things. But I'd like to leave that aside.

Now, we performed a little experiment with Planck's law, too, the first part of the story, that is, the empirical part of the law, the curve fitting. I went over to the faculty club of my university over a period of several weeks and whenever I could, sat down at lunch with someone whom I knew to be a good applied mathematician, might be a physicist, might be a civil engineer, might be a mathematics professor. They're used to me there; they know I sit down in strange places, I don't always eat lunch with my fellow psychologists. So they were not surprised at that.

And then, in the deceitful way that psychologists have, I said to my friend of the day: 'I have some beautiful data. They really are a nice smooth curve, but I can't find a function to fit them. The data are exponential for large values of x; but for small values of x they are linear through the origin. Suppose you could dream up a function which would fit data like that.' I tried that out on eight really first-class people; five of them gave me the answer, all in less than two minutes, and in every case the answer was Planck's law! Now, it wasn't because they previously knew Planck's law and recognized it; because it was characteristic that the civil engineers did not recognize it even after they got it. But it was characteristic that the physicists, after they got it, said: 'Oh, that looks sort of like Planck's law.' But they didn't have that recognition before. Now, alas, they will not get a Nobel prize. They were 80 years too late with their discovery.

But again, we find by very normal processes that discoveries of the first magnitude can be made by bright and well-educated human beings. They had to be well-educated: when I explored the process for a while by which they reached the result, it was clear they were using standard results in mathematics that all people who know those standard results use. For those of you who do know that, I'll simply mention that if you express the answer exponentially as a Taylor series, by subtracting one from both sides you can make it go through the origin and be linear there. And that's what several of them did.

Well, let me leave this part of my story of what we know today about how you can go from raw data to laws describing those data and turn to some other aspects of the scientific adventure. As a matter of fact, when I tell a story something like this to people, they frequently say to me: 'Well, that's all very interesting; but, after all, finding laws from data isn't the real discovery.' They almost always use that phrase, the "real discovery". Well, isn't it? If it isn't the real discovery, then we are going to have to revise our histories of science. We are going to have to make new decisions as to who the heroes are. We are going to have to take the picture of Kepler and Black and Ohm and Planck and Balmer out of the history books or at least take out those parts of their work which was merely finding a law from empirical data. We have to be very careful in German as well as English with talking about something as "the" law. The word "the" is very strong: it implies, first, that something exists and, second, that it is unique. When we say "the" real discovery, we are implying that there is some one act that scientists do that constitutes the real discovery. But I think that's a mistaken picture of what science is. Scientists do many things; and one scientist builds upon, and provides a foundation for what the next scientist does. Experimenters gather new data; theorists use data to arrive at laws and at explanations of those laws in terms of models of the phenomena. When you press and ask, what is "the" real discovery or, as I'd prefer to say it, what are the other aspects of scientific discovery, then there are in fact quite a few. Finding problems is important; and we do regard it as important science if you nail down a problem that turns out to be an important problem. I think it was Helmholtz, for example, who first defined the black-body problem as an important problem. That was a good piece of science. Or was it Kirchhoff?

Finding problems, finding data: data aren't just given. Kepler was lucky enough to have had an employer named Tycho de Brahe who, before he died, produced for Kepler some very fine data. Not all but most of the data Kepler used was data that he inherited. Inventing instruments: you can't do experiments on temperature without having a thermometer. You can't do experiments on elementary particles without having some form of an accelerator. So people invent new instruments. Fahrenheit is in the histories of science partly because he invented a thermometer. Planning experiments on the experimental side - very important part of creativity - is to make some decisions as to what are the right experiments to run in order to solve a problem. Inventing representations: we have to find ways of thinking about the world. Newton had an action-at-a-distance representation. He invented a form of mathematics, the calculus, that enabled him to solve problems using his representation. The Cambridge physicists and mathematicians of the 19th century developed an alternative representation, the so-called field theory; and these representations enable us to solve problems if we can express them in those terms. And, finally, devising explanations, Planck didn't stop with an empirical law that fit the data. He tried to derive further that law in terms of what was known about the underlying behavior of physical objects: Maxwell's laws, the laws of resonators, the laws of thermodynamics, and so on. So there is no "real" discovery. Science is a house of many mansions. That's fortunate, too, because many of us would like to do science; and some of us are very clumsy in the laboratory and break the glass, so we have to become theorists. And others of us are very skillful in the laboratory, but our mathematics is a little weak, so we become experimentalists. So it's very fortunate that there are all of these ways of doing science.

That leads to the question of whether these other aspects of science are perhaps the mysterious ones. Perhaps that's where the creativity is lurking. Perhaps BACON can do things like finding regularities and data; but it requires some other kinds of processes to perform these other functions of science. Let me just give a few examples - this area has been less explored than the area I've talked about - to suggest that the other processes that are part of the whole adventure of science are also understandable, also can be explained, in information processing terms. Let's consider the matter of finding problems. Of course, many of the problems we go to work on are problems already in the literature. One cheap way of finding problems is to ask: 'What problem have people in our field already worried about?'

And then - well, I'll get back to that - if we are cleverer than they (always a dangerous assumption), or if we know something they don't (which might happen), then we may succeed in solving a problem already in the literature. Sometimes, however, it's better to get a new problem, a problem on which lots of bright people haven't already worked, tilled over the ground. But how can you do that? One very important way of getting a new problem is to be surprised. And one very important way of being surprised is to observe phenomena; to go out and look; to use your natural or unnatural - I think it's natural - curiosity. Pasteur, the French biologist, made a very important statement about surprise. He said, 'surprise' - he used the word 'accident', but this is what he was talking about - 'comes to the prepared mind.' Now, why is that?

Well, if something happens, and you don't know anything about it, then you have no reason to be surprised - it might just as well happen as anything else. The only time you can be surprised is when you know enough about the situation to have expectations about what should happen. Now, even those of us who have a little bit of physics here know about gravity and know that objects like our own bodies should stay firmly attached to the floor. If suddenly we began to levitate in this room, most of us would be surprised because we have expectations about the behavior we can expect.

So an English biologist named Fleming, who had the interesting habit of not washing his petri dishes immediately after an experiment but keeping them for a while, one day looked at some petri dishes that he had stored while he went off on vacation. Before washing them, he looked at them once more, and he saw that the bacteria on those petri dishes would die, well, on one petri dish were dying. Now, already his was a prepared mind. I wouldn't know whether bacteria were well or sick; perhaps those of you who come from the medical school would know that. You have to be a prepared mind to know that bacteria are sick. And he also noticed that there was some mould on the edge of the dish. He knew no reason why the presence of that mould or anything else there should cause those bacteria to die; and so he was surprised. And then he also used some rules of thumb, or heuristics, which are quite commonly used by scientists when they are surprised. I can find a half dozen examples of this, maybe more than that. The first thing Fleming did was to try to define the scope of the phenomenon he had discovered. Did it mean all kinds of bacteria or just this particular kind of bacteria? Well, it turned out to be quite a few kinds of bacteria. Did it mean any old kind of mould or just a particular mould, penicillium, which had been on the first dish? Well, it turned out to be only penicillium. But what do you do after you have defined the scope of the phenomena? It was obviously an important phenomenon if this particular kind of mould killed many kinds of bacteria. What do you do when you discover the scope? You try to find the mechanism. And how do you do that? Well, you purify first of all. You take the mould, and you squish them, and you dissolve them in alcohol; you do all sorts of things to see whether you can get something out which is even more powerful than the mould. If you do, you try to purify chemically and find its chemical constitution. And that was accomplished partly by Fleming, partly by Florey and Chain, and led, of course, to penicillin as the first antibiotic that was available to us.

Now you can find the same pattern in the discovery of the tobacco mosaic virus, find the same discovery in the Curies' discovery of uranium: they found a part of their pitchblende which was unusually radioactive and kept purifying and purifying until they had radium, roentgen, and x-rays - many, many examples. So, finding problems by observing phenomena again is not accidental; it happens to people who know a great deal about the phenomena that they are looking at; and it happens because they have some good rules of thumb to apply when they are surprised.

What about finding data, or what about new instruments? Here we have a situation with chickens and eggs. New data spawn new instruments; new instruments spawn new data. The physicists in the United States today are asking the Congress for a few billion dollars - I forget how many - to build a new powerful accelerator. They assure the Congress that, if they get the new accelerator, they will see new particles. I think they are almost certainly right. A powerful new instrument allows us to explore an area that has never been explored before; we'll see new things, just as a ship which goes to a sea that's never been entered by ships before will see new things. And so instruments create, if you want to say that, new data. Now, there really wasn't much in the way of magnetism in the world until we had magnets.

Similarly, data create new instruments. People knew about things being hot and cold for a long time. Then they noticed that very frequently, as you warm things up, they expand; and that was the basic idea that led to the invention of the thermometer. If things expand when they get warmer, why don't we measure the amount of expansion and call that the temperature? That's exactly what happened.

What about planning experiments? And here I want to get back to the computer because here we have some excellent historical evidence, coming from the laboratory of Hans Krebs, the biologist and biochemist from Germany, who later worked in England. Krebs kept his lab notebooks; and they have been very carefully analyzed by a historian of science, Larry Holmes at Yale University; and those data were available to us. What we thought we would do was to see whether we could write another computer program. This one happens to be called "KARKADA", which happens to be the Hindi word which means "Krebs". The person working with me on this project is an Indian student named Kokani. The task of the KARKADA program is to plan experiments. KARKADA plans an experiment; because it can't go into the laboratory and do the experiment, so we tell it how the experiment came out. Then it digests that information, plans a new experiment, and so on and so far. We start KARKADA out in the first instance with a little knowledge of biochemistry - Krebs himself did not have a great deal when he started out - and a research problem.

I should say a word about Krebs's history. Krebs had worked as a post-doctoral student in the laboratory of Otto Warburg; and Otto Warburg in his studies of oxidation in living creatures in the oxygen cycle had invented a powerful new laboratory technique: the method of tissue slices. Until then, if you wanted to study anything in life organisms, you had to take whole organs and keep them operating while you performed the experiments. And whole organs are not very homogeneous tissues, and it was not a very neat situation for experimentation. Otto Warburg found ways in which he could keep slices of tissues on a microscope slide active and do experiments on these tiny slices of tissues - powerful new experimental weapon!

Hans Krebs studied with him and then went off to his own professorship and was set the task of setting up a new laboratory and making his way in the world. Maybe he was only an extraordinary professor, so he still had to worry about getting his professorship. But he had a secret weapon. He had something (not just his own intelligence) other biologists didn't yet have, namely, the method of tissue slices that he had learned in Warburg's laboratory.

And so he set about, first repeating the classical experiments on urea synthesis but using the tissue slice method to see whether he could repeat them; and this involved trying out a lot of amino acids and ammonia to see whether you could get these tissues to produce lots of urea.

And they didn't. They produced a little bit. He tried one amino acid after another; then he got bored of that, and he and his assistant tried a few other experiments that people had done, whether glucose had any effect and so on. Then he went back to the amino acids. One day, his assistant took ornithine, a particular amino acid, off the shelf, ran the standard experiment, and was surprised, surprised because he'd been led to expect certain amounts of urea production; he got a much larger amount. He set about doing exactly what Florey did.

What's the scope of the phenomenon? Is it just ornithine or other things that are chemically similar to it? No, it's just ornithine. Does it depend on how much ornithine? So they try various combinations of ornithine and ammonia. Now, you can use very small quantities of ornithine. The amount of urea is related to the decrease and quantity of ammonia, not to the quantity of ornithine. And so they decide that the ornithine is acting as a catalyst and is recombining and then being released and recombining, etc.

Soon they worked that out. Then they went to the literature to see whether there were any reactions like this. They found a couple of experiments and literature that were helpful, and within a couple of months, they had the reactions worked out.

We give KARKADA the same initial conditions that Krebs had; we give it a Krebs problem; we give it the right to use the tissue-slice method or to plan experiments for using it. We give it a modest knowledge of biochemistry as it was known at that time in the 1930's; and we let it start planning experiments and giving it the results. It follows, in fact, a path almost identical with Hans Krebs's. Once or twice we have to put it back on the track; and after it has its surprise, it does just what Krebs does with his surprise and ends up finding the correct reactions for the synthesis of urea. Now, maybe, that isn't quite surprising, since, of course, we are building the computer program and debugging it with the Krebs example in front of us. That's kind of cheating, you might say. So the real test with rather the same program without changes in the rules of thumb it was using for designing experiments could be moved into other areas of science and still succeed; and it has now succeeded in four - let's say four and a half - cases, counting the original one. We said that it had worked on a 19th century problem that had been quite important at the time of certain syntheses of alcohols. And it solved that satisfactorily. We went back to Krebs and looked at another piece of research he did on the synthesis, on the processes of glutamine metabolism in bodies. We found that KARKADA was also able to do that.

We then gave it the situation that Faraday faced after his first successful experiment in inducing electricity from a magnet. We didn't ask KARKADA to do the first experiment after that. And we found that, in fact, it had planned a sequence of experiments very like the sequence that Faraday actually carried out in the year, or two years, after his discovery.

Now, I could go on and say a little bit about the progress that's been made in understanding the processes of representation and the processes of scientific explanation, that is, going from empirical law down to mechanisms that might tell you why that law would work. Those parts of the theories of scientific discovery are less far along than the ones I have described in more detail here; and I won't try to give you results. Well, I just want to mention one example. There's been a little bit of work done on the atomic theory, atomic hypothesis as it stood around 1800 when Dalton was doing his work. And we find, in fact, that it is very easy to build an inference process which, given the idea - goes back to the Greeks - that what's involved in chemical reactions are little particles - let's call them "atoms"; and given the idea, which is due to Gay-Lussac, that there'll be equal numbers of particles, not atoms but packets of atoms and molecules in equal volumes of gases; given those true ideas, it proves to be rather easy to describe a mechanism which will find the correct formulas for a whole lot of basic chemical reactions and will, in fact, discover atomic weights and be able to distinguish them from molecular weights, which was a problem that confused chemistry for the whole period from about 1800 to 1860 when the problem was solved by Cannizzaro. So we're at least beginning to understand that process and the mechanism embodying the atomic hypothesis which does this. It's also able to reason about things like Mendel's data on his sweet peas: it's able to entertain a kind of chromosome hypothesis or genetic hypothesis about that and is able at least to represent - although we certainly haven't gotten any results - simple phenomena in particle physics. But let me now try to draw this to a close because I'd like to have some opportunity for discussion here.

What, would I argue, this line of research shows us about the human mind, and particularly the human mind engaged in the work of scientific discovery? First, notice that it makes quite clear that inductive as well as deductive processes can be carried out by computers. Some people have been confused about that. Somehow the word "logic" gets associated with computers, and somehow we know that logic is something that does deduction. And what I've been talking about tonight is almost pure induction, going from data to generalizations about that data. So here we have examples of computers doing that. I didn't have time to develop it, but one can also show that there are many processes going on here, which, if humans were doing them, we would call "intuitive". Because the computer has capabilities for recognizing situations and patterns of situations, you can draw on its previous knowledge to make very rapid interpretations of those situations, just as human beings do when they're acting, as we say, intuitively. I think we can claim that we now do have a basic understanding of the symbol manipulation, the inputting, outputting, manipulation, restructuring of symbols that takes place in human problem solving, including creative scientific research as a form of problem solving. And I suppose the most general conclusion one might want to draw from this - although one wouldn't want to stop with the evidence I've given you tonight - is that, like the other phenomena of nature, the human mind is understandable. It's explainable by natural laws.

Some people are a little bit troubled by that assertion; at least it's been my experience in talking on this subject that quite a few people are troubled by that assertion. There are other assertions that have troubled people over the years. At a certain point in history, it was discovered that the earth was not at the center of the universe. In fact, a little later it was discovered that the sun was not at the center of the universe, either. It was out at the arm of a galaxy which was not much at the center of something, of anything, either. Matter of fact, today in many cosmologies we'd have a rather hard time explaining just where the center of the universe is. But certainly we are not added if there is such a center. That troubled people, as we know from the history of Copernicus and Galileo, for quite a while.

I don't think anybody in this room has lost sleep about it recently. We accept that now. We don't think it demeans human beings to know that we are not at the center of the universe. More recently, Darwin came along. He said human beings were not uniquely created as a species; they are created by the same processes of mutation, natural selection that created the other species. And that created a good deal of disturbance. In the United States, it still creates disturbance in some states; and we had to have recent Supreme Court rulings that creationism, the alternative to evolution, doesn't have to be taught in the public schools. So there still are some people who lose sleep about that, probably not many in this room. I haven't found many in Europe; but why not, I don't know. So again, human beings at first felt demeaned by the loss of their uniqueness. And then we realized that that really wasn't important, after all. There's nothing wrong with evolving. Matter of fact, it's kind of progressive, isn't it, to have evolved.

Now, some people - I guess, I haven't been guilty of it tonight - are saying that the human mind is a mechanism (I suppose those of us who are not vitalists prepared for that statement all along), but, moreover, that the things the human mind does in order to think are exactly the things that the computer, ordinary digital computer, is capable of doing; and that again seems to violate our sense of human uniqueness. Of course, we have known all along that, if we have pets, our dogs and cats can think; but they can only think little thoughts. We think big thoughts, and that was supposed to be unique. Now, I don't think we need to be upset about the loss of that aspect of human uniqueness. When a scientist understands something, he or she doesn't cease to admire the beauty of the thing that's understood. I don't know of any astronomers who are incapable of going out at night and enjoying looking at the stars. We can see the state of beauty of the world. For if we're botanists, we can still see the beauty of flowers. So when a scientist understands something, he does not cease to admire its beauty. On the contrary, it acquires a new beauty. The beauty is that, underlying the appearances that we see, that we admire, in the complexity that we see and that we admire, underlying all that, is a real order, is a real orderliness and lawfulness, a set of mechanisms that accounts for the complexity so that we can explain complexity with simplicity.

Simon Stevinus, the great Dutch pioneer in mechanics, published a book about his work, and on the title page of the book he put his family crest. I don't know whether his family really had a crest or whether he invented it. But anyway, he put it on the cover of the book; and he inscribed on that crest a motto in Dutch which, I think, can be translated as: "Wonderful but understandable". And I think that's what we're after in science. We wonder at the capabilities of the human mind which is able to take what appeared to be these great leaps of thought that lead to the world of modern science and the successes that modern science has had in explaining a phenomenon. And the beauty of that process is that underneath it are, in fact, also a set of lawful processes, a set of understandable processes which allow us to know and to grasp how that complexity can arise out of an underlying simplicity.

Prof. Dr. Herbert A. Simon
Carnegie Mellon University
Department of Psychology
Pittsburgh, Pennsylvania 15213, USA

NOTE

1 Paper delivered at the University of Graz.

UNDERSTANDING CREATIVITY

MARGARET A. BODEN

Creativity can be scientifically understood with the help of computational concepts [Boden, 1990/1991]. This may seem surprising, not to say absurd. Most people take for granted that there can be no interesting relation, only utter incompatibility, between computers and creativity. Indeed, most people assume that there will never be any scientific theory of creativity - for how could science possibly explain fundamental novelties? The very notion seems to be a contradiction in terms.

In fact, however, computers and creativity make interesting partners with respect to three rather different sorts of project. One is psychological: understanding human creativity. One is technological: trying to produce computer-creativity. And the third is pragmatic, even educational: using computers to aid one's own creativity.

Human creativity is something of a mystery, not to say a paradox. Artists and scientists rarely know how their original ideas come about. They mention intuition, but cannot say how it works. Nor can they say clearly how creativity can be recognized. One new idea may be *creative*, while another is merely *new*: what's the difference? And how is creativity possible?

Some ideas we call creative are merely novel combinations of familiar ideas. Much poetic imagery, or analogy in science, is of this type. Psychometric measures of creativity usually identify "creative" ideas in terms of their statistical improbability [Sternberg, 1988], and such tests are best-suited to capturing creativity of this "combinational" type. As well as measuring it, the psychologist needs also to be able to explain this sort of creativity.

Imaginative association (whereby vastly different ideas, as well as similar ones, may be related), and the recognition of analogy, can each be illuminated by computational models. Association takes place by means of mental processes comparable to (though often much richer than) the parallel-processing defined within connectionist AI-models, or neural nets [Rumelhart & McClelland, 1986]. The origination of the haunting imagery of Coleridge's poem *The Ancient Mariner*, for instance, can be largely understood in these terms [Boden, 1990/1991, ch. 6]. As for analogy, which often effects a permanent change in one's perception of something, this too has been modelled in AI-terms [Hofstadter *et al.*, in press; see also Holyoak & Thagard, 1989].

Combinational creativity, surprising though its results can be, is less arresting than some other cases of originality. A novel combination of two familiar ideas is something which *did not* happen before. By contrast, some creative ideas surprise, shock, and delight us because it seems that they

J. Götschl (ed.), Revolutionary Changes in Understanding Man and Society, 75–82.

could not have happened before. Relative to our expectations, they are not just improbable, but impossibe.

Combination-theory cannot capture this distinction. An adequate scientific account of creativity needs to be able to do so. Also, it must explain in what sense one impossible idea can be more impossible, more creative, than another.

Non-combinational creativity involves the exploration, and in the most interesting cases the transformation, of conceptual spaces in people's minds. Conceptual spaces are styles of thought. They include ways of writing prose or poetry; genres of sculpture, painting, or music; theories in chemistry, biology, or mathematics; habits of couture; systems of choreography ... in short, any reasonably disciplined way of thinking.

Within a given conceptual space, many thoughts are possible - even if some of them are never thought. Exploration of the space can locate some of these, and find their neighbours and pathways, very often coming up with surprises - in the sense that we had not realized that this space held such a potential. Mozart's music abounds with examples of such ideas, musical gems that fully exploit the possibilities already inherent in a current musical genre. This sort of space-exploration thus includes many valued novelties we regard as creative. It may involve slight changes to relatively superficial dimensions of the space, or the addition of new (also superficial) dimensions. But it involves no fundamental transformations of it. (Mozart was a much less adventurous composer than Haydn, even though he is widely regarded as the greater.)

By contrast, other ideas are impossible, relative to the given space. If you are skilfully writing a limerick, iambic pentameters simply cannot drop from your pen. But if you want to write a new sort of limerick, or a non-limerick somehow grounded in that familiar style, then blank verse could perhaps play a role. The deepest cases of creativity involve someone's thinking something which, with respect to the conceptual space present in their minds, they *could not* have thought before. (It follows that *constraints* are essential to creativity, even if creativity largely consists of overcoming them.)

The obvious next question is how this supposedly impossible idea could possibly come about. And the answer is that the creator must change the pre-existing style in some way. It must be tweaked, or even radically transformed, so that thoughts are now possible which (within that space) were not conceivable before. To understand how this can happen, we must understand clearly what conceptual spaces are, and what sorts of mental processes could explore and modify them.

Styles of thinking are studied by literary critics, by musicologists, and by historians of art, fashion, and science. And they are appreciated by us all. But intuitive appreciation, and even lifelong scholarship, may not make their structure entirely clear. Indeed, the unclarity may be proclaimed as unavoidable by the most scholarly of critics. (An architectural historian said of Frank Lloyd Wright's Prairie Houses, for instance, that their "principle of unity" is "occult" [cited in Koning & Eizenberg, 1981]. By this he seemed to mean forever hidden to human eyes.) Even when a scholar does claim to have discovered a certain style's principle of unity, it may not be stated clearly enough for scientific purposes.

This is the second point at which computational methods can help. Conceptual spaces, and ways of transforming them to produce new ones, can be clearly described by using computational concepts. These concepts are drawn from artificial intelligence (including, but not confined to, connectionist AI), and they enable us to do psychology in a new way. A conceptual space can be thought of as a generative system: a more or less complex set of rules which define the relevant dimensions, and specify ways in which a range of structures (ideas) can be generated.

In this way, the structure of tonal harmony [Longuet-Higgins, 1987], the core of expressiveness in musical performance [Longuet-Higgins, in press], the "grammar" of Prairie Houses [[Koning & Eizenberg, 1981], the inner principle of Palladian villas [Hersey & Freedman, 1992], and spaces of many other sorts [Boden 1990/1991] can be clearly expressed. Then, the power of the computer may be exploited to study various ways of exploring the relevant domain. And methods ("heuristics") for navigating, and even for changing, highly-structured spaces can be examined and compared.

One implication of this account of creativity is that a creative system, whether mind or machine, needs internal "maps" of its own conceptual spaces [Boden, 1990/1991, ch. 4]. These maps, which may exist on many different levels, enable the system to move within the relevant spaces, to test their limits and boundaries, to modify them, and even to transform them. Thus one way, probably the most important way, in which Mozart was different from the rest of us is that his mind contained more richly-detailed maps of music structures, and more ways of negotiating them fruitfully, than other people's.

Computationally-informed work in developmental psychology suggests that young children are unable to vary their skills in flexible and imaginative ways until they have developed internal maps of the skills concerned [Karmiloff-Smith, 1990]. Thus an older child can draw many "funny" animals (two-headed dogs, or centaur-like monsters) at will, whereas a younger child simply *cannot* - even though the younger child's skill at drawing normal animals is no less fluent. Variation of size or shape (of parts and/or whole), repetition or deletion of parts, and mixing of categories (to produce centaurs, for instance), appear in a predictable sequence. The theoretical explanation is that a skill which is fluent may be merely "compiled": it can be run, but not varied. To be varied, it must be redescribed at some higher level (or levels), in terms of specific parameters and subroutines. These "representational redescriptions" not only transform line-drawing (and language [Karmiloff-Smith, 1986]) into a more creative activity, but underlie the development of conscious self-reflection.

What of the second link between machines and creativity? Can computers be creative? That is, can they produce performance of a kind which we would regard as creative if we saw it in human beings? (This is a scientific question. For the purposes of this paper, we may ignore the non-scientific, philosophical, question of whether a computer, *no matter how humanlike its performance*, could "really" be creative [Boden, 1990/1991, ch. 11]).

The answer to this question is "Yes". A number of programs already exist which can *explore* a given space in acceptable ways. For example, a computational grammar has not only shown what the principle of unity of Prairie Houses actually is, but has generated designs for new ones, previously undreamed-of [Koning & Eizenberg, 1981]. Similarly, a program that designs Palladian plans and facades has come up with several designs which closely approximate one of Palladio's own, as well as many more structures in the Palladian style (whose essential nature has been clarified by this computational work) [Hersey & Freedman, 1992]. Some programs can generate thousands of line-drawings in a certain style, pleasing enough (if hung on one's walls) to be spontaneously remarked upon by unsuspecting visitors [McCorduck, 1991]. Others improvize unpredictable melodies, and accompaniments, from a modern-jazz chord sequence [Johnson-Laird, 1988, 1989]. Yet others come up with (occasionally, brand-new) scientific hypotheses [Buchanan et al., 1976; Langley et al., 1987].

The more complex the conceptual space, the more difficult it is to express it in computational terms, so as to produce a computer-creator. The most forbidding spaces of all occur in literature, which draws not only on literary conventions and (largely intuitive) psychological knowledge, but on huge amounts of world-knowledge too. Current computer models of literary creativity (story-writing, for instance) are clumsy and unconvincing, involving infelicities and absurdities of various kinds. Those infelicities help us to appreciate the richness and subtlety of the human mind, but they limit the persuasiveness of story-writing programs as creators.

Story-writing programs can generate plausible stories (expressed in very clumsy language) only where very simple plot-spaces, and very limited world-knowledge, are concerned. One such program produces Aesop-like tales, including a version of "The Fox and the Crow" [Meehan, 1981]. A recent modification of the Aesop-program is more subtle [Turner, 1992]. This system, called MINSTREL, uses a form of analogical ("case-based") reasoning to compose novel stories which are deformations of familiar ones. (Turner calls them "transformations", but the alterations concerned are not sufficiently fundamental to be termed transformations in my sense.) Many superficially distinct plots, and characters, can be generated in this way. Moreover, because MINSTREL distinguishes the author's goals from the goal of the characters, it can solve meta-problems about the story as well as solving problems posed within it. Both these story-telling programs can surprise us, in the sense that their stories are (to some extent) unpredictable. But their literary powers are puny, compared with ours.

Computational definitions of interpersonal concepts used in stories - including cooperation, betrayal, and escape - are also relevant here [Abelson, 1973; Boden, 1977, ch. 4]. Such definitions help us to see the structural affinities between (for example) different sorts of betrayal, such as abandonment and letting-down. These two species of betrayal will involve very different types of character, and very different consequences. For instance, only the strong can abandon and only the weak can be abandoned, whereas anyone can let down, or be let down. A good story-writer, whether human or artificial, needs to have a grasp of these differences. And a writer able to tell a story about any human culture, or even (as in some science-fiction) none, likewise needs to understand how the possibilities inherent in the structure of interpersonal relations vary from group to group.

Naturally, literary programs will continue to improve. They can already generate simple plots that might be of use to an especially unimaginative author of crude types of fiction - such as the lower forms of soap-opera, or of "nurses and doctors" romance. Even if the human author has to embellish the plot, and put the whole thing into acceptable English, some helpful plot-ideas might come from the program (this is relevant to whether computers can help us be creative, a question considered below). But a program capable of writing a genuinely interesting novel or short story is not on the horizon, and probably never will be. The complexity of the conceptual spaces concerned is too great.

Nevertheless, current achievements show that programs *can* apparently be creative, at least in some domains. A few can even *transform* their conceptual habitat, alter their own rules, so that interesting ideas result. For instance, programs using genetic algorithms (see below) have come up with optimal solutions to a number of difficult problems. One such problem concerns accidental leaks in huge gas-pipelines, running across many different countries. A self-transforming, "evolutionary", program can infer the position of the leak on the basis of hourly measurements taken at various points along the pipeline [Goldberg, 1987]. These measurements include the gas-inflow, gas-outflow, inlet-pressure, outlet-pressure, rate of pressure-change, season, time of day, time of year, and temperature. As well as finding the leak, the program can recommend emergency-action (specifying which pumps should be turned off, which valves should be closed etc.).

Many of the ideas generated by current computer-models of creativity were already known to human beings (though not specifically prefigured within the program's initial conceptual space). But the definition of (transformational) creativity adopted within this paper is a psychological one: the ability to come up with an idea which, relative to the pre-existing domain-space in one's mind, one could not have had before. Whether any other person (or system) has already come up with it on an earlier occasion is irrelevant. That is a historical question, not a psychological one.

Moreover, at least one entirely unknown mathematical theorem has been suggested by a computer, whose programmer had never even heard of the branch of mathematics concerned [Lenat, 1983]. The program, using its initial concepts and its exploratory/transformational rules, found its own way into the new mathematical space - by creating that space for itself.

The third way in which computers and orginality are related involves the use of these machines to help our own creativity. A wide range of possibilities for such help exists [Boden, 1994]. For example, consider "Improviser", a program written to help teach people to improvise [Hodgson, 1990, in preparation; Waugh, 1992]. This program knows about various dimensions of the musical space of jazz, and various ways of travelling through it. For instance, it can produce fragments of ascending or descending scales, ensuring that the scale chosen is the one relevant to the harmony at that particular point. It can provide "call" and "reply" over two or more bars. It can replace the current melody-note by another note drawn from the same scale, or provide a chromatic run between this melody-note and the next. It can "cut and paste" a library of melodic and rhythmic patterns, or play fractionally ahead of or behind the beat. And it, and the human user, can vary the frequency with which it does any of these things.

If left to wander through the space by itself, this program will improvise - on a given melody, harmony, and rhythm - by making (random) choices on many dimensions simultaneously. Working in this fashion, it often creates novel musical ideas which professional jazz-musicians find interesting, and may wish to develop in their own playing. Alternatively, the human user can make the program concentrate on one (or more) dimension at a time, and explore it (or them) in a very simple way. This is why it can help jazz-novices, who can focus on the aspect of jazz which is currently causing them difficulty.

Another example of computer-aided creativity concerns "evolutionary" programs. A number of graphic artists, film-animators, and industrial designers, are using computers as genuine partners in their creative quest. The programs in question continually make random changes in their current rules so that entire new forms, new species, of structure result. The human being continually chooses the ones he or she finds most interesting, and the machine concentrates on them.

For instance, an image-generating program [Sims, 1991] uses self-modifying "genetic algorithms" (modelled on biological mutations) to generate new images, or patterns, from pre-existing ones. At each "generation", the selection of the "fittest" examples is done by the programmer - or by someone fortunate enough to be visiting his office while the program is being run. That is, the human being selects the images which are aesthetically pleasing, and these are used to "breed" the next generation. The programmer's aim, in this case, is to provide an interactive graphics-environment, in which human and computer can cooperate in generating otherwise unimaginable images. (At present, this program runs only on the Connection Machine, a massively parallel computer costing 15 million dollars.)

In addition to the unpredictability resulting from randomness, these computer-generated images often cause a deeper form of surprise. It is as if, besides our being unable to predict heads or tails when tossing a coin, the coin sometimes showed a wholly unexpected design. If one compares a parent-image with some of its descendants (or even some of its immediate offspring), one may be amazed by the difference. The change(s) effected to the earlier image on the way to the later one sometimes seem like relatively unadventurous exploration, or tweaking. The colour of what is clearly the same image may have been altered, for instance, and/or the lines may obviously have been blurred. Sometimes, however, one cannot say, merely by looking at the chosen image-pair, how they are related - or if they are related at all. The one appears to be a radical transformation of the other, or even something entirely different.

In sum, there are many intriguing relations between creativity and machines. Computers can sometimes do creative things, and some can help us to do so. Not least, a computational approach can clarify many questions about our own creative powers. It gives the psychologist a way of seeing more clearly into the rich subtleties of the human mind.

Prof. Dr. Margaret Boden
School of Cognitive and Computing Sciences
University of Sussex at Brighton
Falmer, Brighton, BN1 9QH, England

REFERENCES

Abelson, R.P. [1973], "The Structure of Belief Systems", in: R.C. Schank & K.M. Colby (eds.), *Computer Models of Thought and Language*, San Francisco: Freeman, pp. 287-340.
Boden, M.A. [1977], *Artificial Intelligence and Natural Man*, New York: Basic Books, (2nd edn., expanded, 1987).
Boden, M.A. [1990/1991], *The Creative Mind: Myths and Mechanisms*, London: Weidenfeld & Nicolson, 1990; Oxford University Press 1994. Paperback edn. (expanded), London: Abacus 1991.
Boden, M.A. [1994], "Agents and Creativity", *Comm. Assoc. Computing Machinery* (Special Issue on "Agents", ed. D. Riecken), July 1994.

Buchanan, B.G., D.H. Smith, W.C. White, R. Gritter, E.A. Feigenbaum, J. Lederberg, & C. Djerassi. [1976], "Applications of Artificial Intelligence for Chemical Inference: XXII Automatic Rule Formation in Mass Spectrometry By Means of the Meta-Dendral Program", *Journal of the American Chemistry Society 98*, pp. 6168-78.

Goldberg, D. [1987], "Computer-Aided Pipeline Operation Using Genetic Algorithms and Rule Learning. Part I: Genetic Algorithms in Pipeline Optimization", *Engineering with Computers 3*, pp. 35-45.

Hersey, G., & R. Freedman [1992], *Possible Palladian Villas (Plus a Few Instructively Impossible Ones)*, Cambridge, Mass: MIT Press.

Hodgson, P. [1990], *Understanding Computing, Cognition and Creativity*, MSc thesis, University of the West of England.

Hodgson, P. [in preparation], *Modelling Cognition in Creative Musical Improvisation* (provisional title), DPhil thesis, University of Sussex.

Hofstadter, D.R., M. Mitchell, R.M. French, D.J. Chalmers, & D. Moser [in press], *Fluid Concepts and Creative Analogies*, London: Harvester Wheatsheaf.

Holyoak, K.J., & P. Thagard [1989], "Analogical Mapping by Constraint Satisfaction", *Cognitive Science 13*, pp. 295-356.

Johnson-Laird, P.N. [1988], *The Computer and the Mind: An Introduction to Cognitive Science*, London: Fontana.

Johnson-Laird, P.N. [1989], *Jazz Improvisation: A Theory at the Computational Level*, Unpublished working-paper, MRC Applied Psychology Unit, Cambridge.

Karmiloff-Smith, A. [1986], "From Meta-processes to Conscious Access: Evidence from Children's Metalinguistic and Repair Data", *Cognition 23*, pp. 95-147.

Karmiloff-Smith, A. [1990], "Constraints on Representational Change: Evidence from Children's Drawing", *Cognition 34*, pp. 57-83.

Koning, H., & J. Eizenberg. [1981], "The Language of the Prairie: Frank Lloyd Wright's Prairie Houses", *Environment and Planning B, 8*, pp. 295-323.

Langley, P., H.A. Simon, G.L. Bradshaw, & J.M. Zytkow [1987], *Scientific Discovery: Computational Explorations of the Creative Process*, Cambridge, Mass.: MIT Press.

Lenat, D.B. [1983], "The Role of Heuristics in Learning by Discovery: Three Case Studies", in: R.S. Michalski, J.G. Carbonell, & T.M. Mitchell (eds.), *Machine Learning: An Artificial Intelligence Approach*, Palo Alto, Calif.: Tioga.

Longuet-Higgings, H. C. [1987], *Mental Processes: Studies in Cognitive Science*, Cambridge, Mass.: MIT Press.

Longuet-Higgings, H. C. [in press], "Artificial Intelligence and Musical Cognition", *Phil. Trans. Royal Society*, London: Series A (Special Issue on "Artificial Intelligence and the Mind: New Breakthroughs or Dead-Ends?").

McCorduck, P. [1991], *Aaron's Code*, San Francisco: W. H. Freeman.

Meehan, J. [1981], "TALE-SPIN", in: R.C. Schank & J. Riesbeck (eds.), *Inside Computer Understanding: Five Programs Plus Miniatures*, Hillsdale, NJ: Erlbaum, pp. 197-226.

Rumelhart, D.E. & J.L. McClelland [1986], *Parallel Distributed Processing: Explorations in the Microstructure of Cognition*, 2 vols, Cambridge Mass.: MIT Press.

Sims, K. [1991], "Artificial Evolution for Computer Graphics", *Computer Graphics 25* (no. 4), pp. 319-328.

Sternberg, R.J. (ed.) [1988], *The Nature of Creativity: Contemporary Psychological Perspectives*, Cambridge: Cambridge University Press.

Turner, S. [1992], MINSTREL: A Model of Story-Telling and Creativity, *Technical note* UCIA-AI-17-92, Los Angeles: AI Laboratory, University of California at Los Angeles.

Waugh, I. [1992], "Improviser", *Music Technology*, September 1992, pp. 70-73.

THE ROLE OF SIMULATION MODELS IN THE COGNITIVE SCIENCES

ARTURO CARSETTI

As is well-known, we normally distinguish at the level of biological languages, from a general point of view, between competence and performance and between surface structures and deep structures.

At competence level it is necessary to consider the presence of a series of generative abilities and potentialities, of processes that cannot be determined, in a complete way, through the explicit definition of a finite, static system of given rules: namely of recursive rules. In this sense, one must explain how to describe, at a formal level, the characteristics and the nature of the instruments of invention and "guide" which have a central role in bringing to light the generative abilities that lie at a potential level and in shaping and recognizing new linguistic structures as well as new modalities for the description of processes. These conceptual instruments are the tools which an owner of a natural language uses every day when he articulates his cognitive activity and which gradually determine the development of the language itself.

A careful analysis of the distinction between competence and performance has, in recent years, led J. Nelson to a deeper reading of the mesh of relationships existing between Turing machines and natural languages.[1]

From an abstract point of view, a TM may be seen as a program machine which can perform computations on its own program and that can even reach conditions of universality: that is, as a machine that has the capacity to simulate the behaviour of any other machine including itself. This means that it might calculate any function which, according to Church's thesis, can be computed by mechanical means. It is precisely from this thesis stating that "all effectively calculable functions are recursive functions", that Nelson in 1982 proposed a "strong" version which is particularly useful for a deeper description of the notion of competence. Nelson suggests extending Church's thesis by substituting the concept of effectively calculable function with that of effective process, and by claiming that every effective process is a recursive process and vice-versa. The chief purpose of this extension is to account for the possibility of discovering, in the future, particular languages and axiomatic systems within which it would be formally possible to represent classes of effective processes which it has not as yet been possible to express analytically in accordance with the conceptual tools of the classical recursion theory.

Extending Nelson's primitive idea we could say, in this sense, that the term "recursive" will indicate all the equivalent "ways" (but specified according to the logical paths and the constraints which are necessary at

J. Götschl (ed.), Revolutionary Changes in Understanding Man and Society, 83–91.

times: paths and constraints which cannot be directly expressed, at the beginning, in a combinatory manner, within the conceptual frame of the first-order logic and which determine the type of effective realization of the aforesaid "ways") of developing and explaining, at surface level, the intuitive "idea" of effective process which turns out to be the real primitive fact we must analyze.

This kind of conception presupposes the possibility of proceeding towards a gradual introduction, with respect to the classical tools utilized by the recursion theory, of non-standard methods of analysis and conceptual construction, outlining the traits of an "intensional" dimension of the computation theory.

In this way specific kinds of effective processes until now not formally definable, will finally result traitable and simulable (even if according to precise approximation procedures and coupling mechanisms).

The arising new textures of recursive rules could be regarded as the first steps by which it is possible to canalize and specify, utilizing heuristic, semi-algorithmic and self-organizing methods, the correct development of the potential faculties hidden in the deep structure of a "creative" processor.[2] A processor, that is, which should be represented, from an abstract point of view, as a "biological" processing system with a fairly high degree of complexity, endowed with a deep generative basis, with reflexive operators, with representational and ideational areas and with a knowledge level: as a system which operates in accordance with the modules of an autopoietic system.

Such a development and enlargement, with respect to the range of application of the classical tools of computation theory, should be considered as one of the main conditions for the possible growth of the same stable texture of the processing system's computations. This growth appears necessarily connected to a continuous restructuring of the underlying informational processes, to a progressive expression of new grammatical inference procedures and to a successive assimilation of the information that lives in the external world.

The "biological" computer, in this sense, should be considered as a processing system essentially characterized by the existence of a continuous coupling between its deep generative basis and the representational and "neuronal" level.

The neural level, by means of simulation activities and inventive procedures, as these latter articulate themselves in the "landscape" represented by the effective articulation of selective pressures, can prime, in a mediate way, (by means of patterns of selection through elimination which act on the mechanisms of regulation of the constraints which live at the deep level) the firing of hidden connections, the autonomous evolution of new forms of action.

This evolution can, in turn, determine, at surface level, the imposition of specific constraints, the appearance of new textures of rules and the birth of particular assimilation processes.

So, if we consider the programming languages as modalities for the description of processes, the development and the expression we are speaking of could be considered as the successive disclosure of extended parts of possible "silent" processes-programs, of pre-algorithmic paths never envisaged before. Of structures, that is, which potentially exist within the processing system (hidden in the deep structure of the processor) and which have never before been revealed and "constructed" effectively as programs because of the absence of the conditions needed for their revelation or for their actual construction.

We have to stress, however, once more, that it is only the possible invention at surface level of the aforesaid new tools of conceptual construction that can allow the discovery and the development of these hidden structures.

At the same time we must underline that the aforesaid invention doesn´t operate in a Lamarckian way, but only with reference to the progressive articulation of selective pressures, an articulation which results essentially coupled with the simulation activity. It is only in this way that an effective dialogue between deep information and surface reality can arise. Contemporarily there will be the breaking up, on the basis of selective activities, of the old constraints which determined the constitution of certain "replicators" and the successive appearance, along the first articulation of new self-organizing constraints, of new replication structures, of new forms of expression of the deep information.

So the inventive process, the connected realization of new simulation models of a higher-order level, cannot be simply seen as a coded description of a given system according to von Neumann´s theory of self-reproducing automata, whereby a coded description that the processing systems uses as a sort of working model of itself which, at the same time, need not be decoded.

On the contrary, this inventive process should be considered, first of all, as an inner guide for the derepression processes which live inside the system. One must not add from outside new rules to the model-system, but make it as an autopoietic model-system able to interact with the dynamics that it tries to describe in order to outline the birth of new principles of coagulum able to guide, in a mediate way, a "creative" development of the deep basis which subtends such dynamics.

In other words, if we consider man as a true biological creative processor, the outlining by man of such a simulation system doesn´t signify the construction of a simple, static representation of himself or himself in the world. The model constructed must, in its turn, serve, at the human level, to derepress man´s deep reality, to establish, that is, the necessary conditions for the birth of a constrained dialogue with the information living in the deep dimension. This dialogue can take place only by means of the successive specification of well-defined selective processes which concern, within the more general frame of the selective pressures present at the phenotypic level, the same simulation model. In this way, the model will progressively allow the realization of an epigenetic growth of an objective nature.

According to this point of view, the representations relative to the model (the abstract model itself considered as a representation) don´t have any substantial character for man, but only a functional one. These representations will progressively become embedded in the dynamics, in the epigenetic growth that is taking place according to the selective constraints imposed by the enviroment.

This fact doesn´t prevent the model (if it turns out to be a "true" model) from attaining conditions of real autonomy (even if of a limited autonomy). In this way, the model will show itself as an objective reality and not only as a subjective projection of the neural apparatus.

Actually the model (if successful) can induce, as we said, in a mediate way, the development of a specific coagulum action able to canalize generative principles lying at the deep level. To the extent that the model will succeed in acting as an indirect tool for the realization of such a development, it will reveal itself as truly autonomous. This fact will allow the model to elaborate an objective representation of itself. A form of simulation of simulation. In this way it will also allow the actual realization of a form of creativity on behalf of the processor, of a form of artificial life, which results in being life and intelligence: intelligence considered as an ideal Form, as the recognition of a coded "cipher" on which the self-organizing processes, living at the deep levels of the source, can finally graft, in an indirect way, a coherent expression of their potentialities. According to this point of view, an autopoietic simulation model should live in strict connection to a natural and semantic dimension. A natural dimension, however, that cannot fully express itself except through the utilization of formal methods, ideal representations and particular systems of rules of production.

In this way, the real mirror represented by the design of an adequate simulation model will result to be an integral part of a larger process of natural evolution. From a general point of view, a natural creative dimension cannot develop all its potentialities, once a certain level of complexity has been reached, without the exploitation of the inner growth of the mirror represented by the simulation model. A model which, to be a real mirror and an effective canalization factor at cognitive level, should, however, show itself, as we have said, to be increasingly autonomous, closer and closer to the complexity of the biological roots of the natural dimension and increasingly creative, even though within the effective limits of a coupled system.

The autonomy achieved by the model, within the aforesaid limits, guarantees that the neural growth partially determined, at the human level, by the coagulum activity expressed by the system has been obtained according to objective modules.

The coagulum activity which canalizes the generative principles lying in the deep semantic dimension, manifests itself indirectly and is achieved through the formation of attractors (of sets of attractors) which constitute the fixed points of the dynamics taking place, of the occurring epigenetic growth. So the results of this activity can be seen, at the same time, as an objective representation of the creativity and the partial autonomy reached by the model and as a reliable evidence of the realized unfolding of the potentialities hidden in the natural dimension.

The model, in this way, must show itself to be one that is unable to survive away from the paths of the natural dimension and as one that can become an integral part of its evolution in accordance with precise biological constraints. So , once the model has been separated from the creative flux existing at the natural level, it will crystalize and become a simple automatism. On the contrary, if the mirror-model is connected to the creative deep dimension in a continuous way, it might be able to grow with it and continually renew its own roots according to the successive steps of the realized creative development.

In this sense, the model can reveal itself as a form of "artificial" but living intelligence and, at the same time, as a form of "true" cognition.

Now if we take into consideration, extending and "rewriting" Church´s thesis in a coordinate way, the following two theses formulated, respectively, by Turing and Post:

1) every humanly effective computation procedure can be simulated by a Turing machine
2) any law we become completely conscious of can be mechanically or artificially constructed

we can easily remark that even natural man, if he succeeds in identifying himself completely with a particular given formalism, necessarily determines a form of crystalization, allowing, in this way, the progressive identification of his natural dimension with some form of automatism. According to this point of view, we can realize that the invention and the construction of a mirror-model in an unended way, results in being, for a cognitive system, a real attempt to escape from the dangers of crystalization in order to reveal, step by step, all its potentialities. In this attempt is hidden the secret of the autonomy of the cognitive system, the secret of its freedom but also, at the same time, of the possible loss of its inner consistency. The construction of the model, in truth, doesn´t constitute only a mental representation or a form of awareness: it represents the necessary condition required in order that the cognitive system can reproduce itself, not in von Neumann´s sense, but according to a creative development characterized by the existence of selective procedures and by a specific coupling with the environment.

From a technical point of view, the creative processor cannot outline a simulation model of itself if not with reference to the articulation of specific higher-order logical and intensional procedures. (Without taking, in other words, into consideration the holistic traits proper to the global process that characterizes the flux of information back and forth existing between the generative basis and the "neural" level and between the same system and its model).

Actually the processor, considered as an abstract processing system with a high degree of complexity, cannot prove its consistency by methods of proof which are restricted enough to be represented within the system itself. On the other hand, the simulation model should be able to prove theorems concerning its proper limits[3] and, at the same time, to outline a possible extension of Hilbert´s finitary standpoint by establishing some of the necessary conditions for the utilization of certain abstract concepts in addition to the merely combinatorial concepts referring to symbols; by establishing, that is, specific bases for the utilization of concepts that do not comprise properties or relations of concrete objects, but which are concerned with higher-order "thought" constructions.

Precisely the simulation model should be able to outline, at a "schematic level", some of the necessary conditions for an increase of the complexity of the inner relational structures and of the patterns of symbolic manipulations capable of priming, at the human neural level, specific processes of epigenetic growth. So it should be able to simulate complex "scenarios" and to build a laboratory of the "possible" anchored to the identification and exploration of non standard models: to the exploration, that is, of those paths which outline the changing texture of virtual worlds, of new intended models.

Such an exploration should, evidently, be considered strictly connected to the work of selection; to be precise, it should be considered as a tool for the revelation of new possible paths of natural selection. Only to the extent of this type of revelation, could the laboratory of the "possible" act, in an objective way, as an indirect matrix and as a possible primer for the aggregation of new mechanisms of breaking off and binding of the constraints and for an informational reorganization at the level of the deep generative basis. In other words, it is only the coupled work of the selection and the simulation that can establish the conditions for the realization of an indirect action of guide for the depression processes of the information that lives at the deep levels of the source.

These processes will determine, in turn, the appearance of new patterns of possible constraints that will graft themselves on the tissue of the inventive activities determining the emergence of new forms of expression and new lines of action.

From a semantical point of view, the simulation activity induces the design of a continous cycle of addition of worlds to worlds and determines a holistic and modal dimension of the coupled system. Only a higher-order logic can explain, from a formal point of view, in a complete way, the diachronical character of this dimension. It is only with reference to this particular logical dimension that, as we have said, the flux of information can be progressively specified according to precise forms of coagulum and selection able to canalize the revelation of the particular potentialities present in the source.

Along this line of thought we have to stress two major points:

1) According to Post´s thesis, if we are able to outline, within the boundaries of specific mathematical approximation methods, an autopoietic model capable of reflecting our intellectual functioning and of acting as a true derepression factor, this once realized, will reveal itself as a processor able to prove theorems concerning its proper limits (the limits, that is, of the formal system represented by the model). At the same time, the model will appear to be characterized by a continuous informational and selective exchange with human brains. From this double informational coupling it will be able to extract, by means of specific patterns of constrained assimilation, the lines for its possible evolution. The model, in other words, will reveal itself as an autonomous system linked to man by a selective rather than an instructive relationship. In this sense, the awareness obtained of its own limits represents for the model the first step to be taken in order to achieve "true" autonomy.
2) On the other hand, the man that examines the secrets of the biological nature of his brain can find an explication of the riddle represented by his own nature only by using simulation models, models that are becoming now more and more autonomous and complex, characterized, in perspective, by the existence of biochemical and biological hardware. The progressive autonomy of these autopoietic simulation models appears to allow and guarantee their successful utilization in guiding the informational reorganization processes as the level of the source.

With regard to this last consideration, we have again to stress that, as far as the model results to be a real derepression factor, it will be progressively embedded (at the human level) in the arising epigenetic structures.

In this way, the representational character of the mirror will disappear. The only thing existing and acting, when conditions of real autonomy are reached, is the working dynamics. The autonomy of the model-system, its state of being as "res", enables man to have an objective representation of his own creativity. This representation is embedded in the real neural growth in order to guide the unfolding taking place in an objective manner. Hence the arising multiplicity of autonomous centers which, however, result in being indissolubly coupled.

So, if we bluntly take away the representations, the intellectual game between the possible and the actual, we prevent the effective revelation of the deep information content. The model, the representational mirror, is used not to describe objectively in the neural structures an invariant and independent reality, but to guide the epigenetic growth in order to elaborate new successful schemes of intelligent action.

More precisely, it is used to canalize the embodiment of the deep generative principles in the arising neural growth in such a way that new Gestalten are born at the level of the cognitive network.

These Gestalten will then be used by the observer as tools of objective knowledge of the world. They can be considered as both analytical forms of reading and of generation of natural processes.

On the other hand, we have again to underline that to the extent that the model will reveal itself as a real factor of canalization, it will result truly autonomous and in possession of a form of artificial "intelligence" (of a form of "intelligent life"). In this sense, a particular embodiment of the model (if it turns out coupled in a continuous way with the successive articulation of the natural dimension) in a highly complex biochemical apparatus will be possible.

The realized informational coupling will guarantee the correct "passage" of the coded pattern of the emergent new life. The resulting embodiment will reveal itself, in turn, as a real support (which acts as an external working model) for guiding the progressive unfolding of the generative principle, hidden in the originary source.[4]

According to this general point of view we can say, therefore, at the end of these short observations, that the construction of autopoietic simulation models, on the one hand, and the process of discovery and unfolding of deep biological information, on the other hand, represent, at this time, the two interconnected paths consciously pursued by natural man in order to reach a real expression of his creativity and of his ability to generate the truth in a coherent and objective way. This global process essentially finds, at the moment, its own aim in the gradual attempt to construct particular forms of artificial life: that is, in the tentative renewal on behalf of man, by means of simulation models, of the same roots of natural life. It results in being a difficult, mysterious and risk process within which, in any case, man can find an effective measure of his biological freedom and of his intellectual responsibility.

Prof. Dr. Arturo Carsetti
Via Lariana 7
00199 Rome, Italy

NOTES

1 See Nelson, J., *The Logic of Mind*, Reidel, Dordrecht 1982.
2 See Carsetti, A., "Natural Intelligence and Artificial Intelligence", in: *Intelligent Information Systems for the Information Society*, (B.C. Brooked ed.), Elsevier, Amsterdam 1986.
3 See for more details: Webb, I.C., *Mechanism, Mentalism and Metamathematics*, Dordrecht 1980, Ch. IV.
4 Cfr.: *From Complexity to the Emergence of Meaning*, (A Carsetti ed.), Rome (in press).

THE MIND AND COMPUTATION[1]

JOHN R. SEARLE

I.

In any intellectual era, there are certain views that are widely held that are, nonetheless, implausible to the point of being out of the question, and even out of the question given the state of knowledge of the particular era in which they flourish. Subsequent generations find it almost incredible that their great grandparents could have believed anything so far-fetched. In the philosophy of a hundred years or so ago, I believe an example would be the absolute idealism of the Hegelian variety that flourished at the end of the nineteenth century. It is hard now even to recover the mode of sensibility which made it seem plausible, even true, that everything in the world is really a set of blips on some great cosmic mind. A more recent example of widespread absurdity is the behaviorism that flourished in psychology and philosophy only a generation ago. The idea that there is really nothing to your mind except your bodily movements seems to me so implausible that I can't easily imagine how people could have believed it. However, even behaviorism isn't quite totally dead.

I think one of the hardest tasks in philosophy is to identify contemporaneous forms of absurdity that are widely held and then to expose their absurdity. A form of absurdity, widely held today, that is just as implausible as absolute idealism or the behaviorist theory of mind is the computational theory of the mind - the idea that the brain is a computer and that the mind, including the conscious aspects of the mind, is just a formal program implemented by the computer. This view is widely held today by otherwise intelligent people. I have argued against this form of absurdity elswhere, and in this article I want to continue my campaign to expose it.

Before discussing computationalism explicitly, I want to present what I think is the correct view of the relation of the brain to the mind. I want everything I say in the next few paragraphs to seem blazingly obvious, absolutely platitudinous, given what we know about how the world works in general and about biology in particular. Here goes:

The brain is a bodily organ like any other. It is composed of cells. It is encased in a skeletal structure, the cranium, and it is nourished, like other organs, by way of blood carried through the arteries and veins, due to the pumping action of the heart. The special features of the brain are its capacity to control a large number of aspects of our biological life, such as breathing, digestion, etc., and from our point of view as philosophers, its capacity to produce all of the enormous variety of our mental life, especially

J. Götschl (ed.), Revolutionary Changes in Understanding Man and Society, 93–105.

consciousness. Consciousness is a higher level feature of the brain caused by the behavior of lower level microanatomical elements, especially neurons and synapses. We don't know the details of how the brain works, but we know that as a matter of fact, it does work. It works to produce not only all of our consciousness, but all the other aspects of our mental life such as unconscious mental processes and memory.

We can summarize the relation of the mind to the brain with two simple slogans:

1) Brains cause minds.
2) Minds are higher level features of brains.

If we fully appreciated the consequences of these two rather obvious points a very large number of false views in the philosophy of mind would simply disappear. However, I am getting ahead of my story, so now let us turn to the computational conception of mind.

II.

There is no point in trying to expose the absurdity of the computational theory of the mind if I cannot at first make it seem immensely plausible. I want to present now the strongest case I possibly can for the computational vision of the mind. Here goes:

The brain is above all an information processing system. It takes in information in the form of sensory stimuli at peripheral nerve endings. By means of transduction it converts this information into a series of computational states. It then processes this information in a way that enables the organism to cope with the environment and the coping behavior of the agent is a matter of transduction at the outputs of the information processing system into mechanisms for motor behavior. To take an obvious sort of example, when I go into a restaurant, I take in information in the form of sensory input data about the restaurant, the food, the menu, the prices, etc. I feed this information into my data base which includes not only personal information such as the fact that I am hungry, but also social information such as what constitutes appropriate behavior in restaurants. I then output information in the form of orders to the waiter to bring me such and such food, and in the form of output behavior such as eating the food and paying for it.

Now one thing we know from the past fifty years of studying information processing systems, is that they admit of different levels of description. To take a simple sort of case, the thermostat that turns on my furnace at home takes in information about the temperature in the room. It matches this information against the thermostat setting and then it turns on or turns off the furnace as is appropriate to the setting. In this information processing description, I didn't need to say anything about the lower level of description as to how the information is realized in or implemented in the hardware. It doesn't matter how information processing systems are implemented in the hardware provided only that they process the right information in the right way. The description of something as a thermostat is already sufficient to tell us the right level of description for its information processing functions. The implementation in the indefinitely large range of hardwares at which you could do the relevant information processing is totally irrelevant to the functional level in which we are really interested.

The idea of information processing, and the idea of an information processing system as admitting distinctions between the information processing level of description and the hardware implementation level of description, are the first steps on the road to the computational theory of the mind. To get all the way we need four more ideas, each an idea of great depth, but each admitting of a simple presentation for our present purposes.

First we need the idea of a Turing machine. A Turing machine is a device which does computation by breaking down complex computational tasks into their simple components. Indeed a Turing machine is so simple that it can perform only four basic operations. It can erase a zero and print a one. It can erase a one and print a zero, it can move one square to the left. It can move one square to the right. It does these operations according to a program, and a program is a set of instructions of the form "If condition C obtains, perform act A".

The importance of the idea of a Turing machine for our present discussion is this: with certain qualifications I will get to in a minute, what we ordinarily think of as digital computers are Turing machines.

A second fundamental idea is Church's Thesis. A fascinating feature about information processing is that whenever the information processed can be given a precise formulation, and when the processing in question can be stated in the form of a precise algorithm, there is some Turing machine that can implement the algorithm. Every algorithmically solvable problem is solvable on a Turing machine. And this point is not a piece of philosophical speculation, it is on the contrary, one of the most important mathematical results of the twentieth century. As I said, it is called Church's Thesis, or sometimes the Church-Turing Thesis. And for our present purposes the importance of it is this: we should think of the computer program as a function in the mathematician's sense. It maps an input onto an output. What Church's thesis tells us is that any computable function can be computed by a Turing machine.

Now, there is third idea that is also stunning in its implications. It is this: there is a universal Turing machine such that it can simulate the behavior of any other Turing machine. This result is called Turing's theorem.

When I say such things as "There is a universal Turing machine such that..." or "Turing machines can do such and such..." I don't mean that there is an actual hunk of machinery somewhere on which is written "Turing Machine". The idea of a Turing machine is an abstract mathematical idea; it does not refer to an actual piece of machinery. However another remarkable thing has happened: we have built actual pieces of machinery that can implement Turing machine programs, and for practical purposes they are Turing machines. They have features that the ideal Turing machine does not have. For example, they require electricity, they break down, and they do not have infinite storage capacity; but for all practical purposes, the household computer is a universal Turing machine.

Now, what made these three results so exciting, the idea of a Turing machine, Church's thesis and Turing's Theorem is the following thought: suppose the brain is a universal Turing machine? If you think about it, it's hard to see how the brain could be anything other than a Universal Turing machine, except of course for such irrelevant limitations as those we found for household computers. The brain has a finite amount of information and eventually it dies, neither of which are true of universal Turing machines, which have a purely abstract mathematical existence.

The fourth idea that supports the computational theory of the mind is an empirical test for determining when computers have the same mental states as human beings. Turing actually gave the computational theory of the mind empirical content when he proposed an actual test we could perform on computer simulations. It is called "the Turing Test", and without too much oversimplification we can summarize it by saying that if you program the computer so that it can perform tasks in a way that is indistinguishable from the performance of an intelligent human being, then the computer has proved it has human intelligence.

Just to summarize these points: we now have four ideas which turn the idea of the brain as an information processing system into the idea that the brain is a digital computer and that the mind is a computer program. These four ideas are so powerful that they transform the study of the mind from philosophical speculation and psychological confusion into a genuine cognitive science. They are the ideas of a Turing machine, Church's Thesis, Turing's Theorem, and the Turing Test.

Now, once you accept this hypothesis, then suddenly all of our problems have a possible solution. First, we don't need to worry about the details about how the brain works as a wet, messy biological system. The brain is just the hardware in which we happen to be implemented. Any old hardware would do provided that it is sufficiently rich and stable to carry the programs that our brain, in fact, carries. Secondly, after two thousand five hundred years of fruitless philosophical debate, we now have an obvious solution to the mind-body problem. The mind is to the brain as the program is to the hardware.

For ordinary computers we know exactly how the program is implemented in the hardware, so in principle at least we can come to know exactly how the mind is implemented in the brain. And as Zenon Pylyshyn puts it, the computational theory of the mind is the only non-question begging answer that has ever been given to the question 'How is it possible that the semantic contents in our minds can actually affect the physical behavior of our bodies?' The answer is that the mental contents of our minds are computational symbols implemented in the physical system of the brain.

Third, we have an empirical test that turns this whole hypothesis into a genuine empirical science. The Turing test enables us to find out when we are in fact producing actual mental states on a digital computer of the kind that you can buy in a store: if the machine can behave in such a way that its output is indistinguishable from that of a human; if, for example, it can behave in a way such that experts cannot tell the difference between the machine and the human; then the machine has the same cognitive capacities as a human. If the machine gives the right answers to questions in Chinese, then it understands Chinese in exactly the same sense as the human understands Chinese.

Fourth, and finally, we now have a research project. The research project is to design programs that will pass the Turing test and thus literally have understanding. Once we have achieved this understanding, we then get empirical psychologists to test, by such methods as reaction times, whether or not the programs for cognition that we have designed in artificial intelligence are really the same as the programs that are running on a human brain. So at long last, psychology, the philosophy of mind, linguistics and no doubt, other disciplines such as anthropology, sociology and economics are finally on a scientific foundation.

In recent years this project has been enriched and varied with certain additions. Perhaps we should not think of the brain as a digital linear computer of the Von Neumann variety but rather think of it as a series of computational systems operating in parallel and in an interactive fashion, somewhat like the ways in which the neuronal system in the brain actually works. This new variation called, 'connectionism', 'neuronal net modelling' or in some versions 'parallel distributed processing' is seen by some as a rival to the traditional digital computer model of the mind, but it need not be so taken. Some people think of the mind as a Von Neumann style computation system implemented in a connectionist or PDP hardware system. But for both views again, it is important to see that the actual physics, chemistry, and biochemistry don't matter at all to the actual working of the mind, they just happen to be the form in which the hardware implements the programs. Whether the programs are linear, Von Neumann programs, or parallel distributed programs, the brain is merely the physical implementation.

A final feature to notice about the computational theory of the mind is that it isn't just a fantasy or a distant research program. It is actually being carried out right now. Brilliant work has been done in the study of syntax by Chomsky, and a whole school inspired by Chomsky, which is computational both in its inspiration and in its structure. Chomsky's rules of syntax are purely formal or, as he himself says, "computational". Furthermore, the computational theories of vision, such as David Marr's classic work, *Vision*,[2] are at last giving us a scientifc account of vision by showing how the visual system takes in information about a two dimensional visual array on the retina and prints out information in the form of a three dimensional description of the external world.

That, in a nutshell, is the computational vision of the mind and of the relationship between the mind and the brain. Does it seem convincing to you? Does it seem plausible to you? If I have not made it seem at least plausible, then I have not done a good job of presenting it or you are not a fully socialized member of late twentieth century intellectual culture.

III.

When I first encountered the computational theory of the mind, it seemed to me there was an obvious one sentence refutation: computational states can't be equivalent to or identical with mental states because computational states are defined purely formally or syntactically, in terms, for example, of sequences of zeroes and ones, whereas mental states like wishing you had a cold beer or believing that Clinton will be unable to solve the problem of the deficit, are states that have more than a syntax, they have an actual mental content, they have a semantics. In short, the failure of computationalism can be stated in four words: syntax is not semantics.

I demonstrated this point over a dozen years ago with a simple thought experiment. Just imagine that you are locked in a room manipulating a lot of symbols in a language you don't understand, say Chinese, for example. Now, you can manipulate the symbols according to syntactical rules, and you can, in principle at least, have a program that enables you to give the right symbols as output in response to the right symbols as input. You can pass the Turing test for understanding Chinese. But, if like me, you know absolutely no Chinese to start with, you still do not understand Chinese. And if you do not understand Chinese on the basis of implementing the program that passes the Turing test, then neither does any other computational system understand Chinese solely on that basis, because no other computational system has anything that you do not have. All it has, like you, is a syntax; and syntax is not semantics.

That argument is decisive, but in fact, I now believe that it concedes too much to the computational theory of the mind. So let's explore the matter a little bit further. Some people might agree, and many do, that the Chinese Room Argument refutes one version of the computational theory of the mind, what I call Strong Artificial Intelligence, the view that having a program is sufficient for having a mind. But they still think that having the right program might be necessary, might be an essential explanatory part of having a mind. In a word, they think that though mental states do indeed have a semantic content, and hence are not reducible to their syntax, nonetheless, they all also have a syntactical structure; and this syntactical structure constitutes the essential explanatory level in psychology, because mental processes are computational processes operating over the syntactical or computational structure of mental states. That's the basic idea that survives in computationalism after the refutation by the Chinese Room Argument. This idea, for example, is implicit in Fodor's conception of the Language of Thought.

Now, just to have another name to distinguish this view from Strong AI and Weak AI, I call it Cognitivism. If Cognitivism survives the Chinese Room Argument, then what shall we say about it? The point is that the Cognitivists can concede the Chinese Room Argument but that's unimportant to their account. On their view, minds do have a semantics so the program is not constitutive of the mind, but the essential explanatory level in Cognitive Science is syntactical. Semantics, or mental content, though it exists, is not essential to the Cognitivists' explanation of the mind - syntax is essential.

Both versions of the computational theory of the mind, both Stong AI and Cognitivism, ought to worry us initially because they say that where the mind is concerned, the brain does not really matter. Any computationally equivalent hardware would do as well. This ought to worry us, because they are saying that unlike any other biological capacity, such as the capacities for digestion or growth, biology simply does not matter to the mind. The mind happens to have a biological implementation in the brain, but any other computationally equivalent device made out of anything, beer cans, transistors, water pipes, cogs in wheels - you name it - anything at all could do exactly what our brains do. Let us think about this point for a moment. People who accept the computational theory of the mind are delighted with the feature they call "multiple realizability": the same computational system can be realized in a multiplicity of computational hardwares. Philip Johnson-Laird, an enthusiast for the computational theory of the mind, writes that a system made out of levers with cogs and wheels could implement the relevant computer programs.[3] Zenon Pylyshyn, another enthusiast, writes that a system made out of pigeons pecking could be a digital computer.[4] Even more spectacularly, Ned Block writes that a system made out of cats and mice and cheese, a system where a gate opens and the mouse runs out to get the cheese and the cat pounces on the mouse - that such a system could also be a digital computer. One element would be a zero, and another element would be a one. The point, Block tells us, is that the physical realization doesn't matter provided that the systems are computationally equivalent.[5] The authors I have cited are delighted with this result but I think they ought to be worried about it. If I had a philosophical theory that had the results that you could make a brain that does exactly what my brain does, out of, for example, cats and mice and cheese - you would just need an awful lot of cats and mice and cheese - I would be very seriously worried about that result.

So, with this in mind, I began to probe the literature just a little more closely. If somebody says that the brain is really a computer and the mind is essentially a computer program manipulating formal symbols according to computational rules, then it seems to me that we had better find precise definitions for these notions. It was at this point that I made a surprising discovery. As far as I can tell from the literature, there is no universally agreed on, precise definition of any of these notions. We have, on the one hand, a very powerful mathematics that includes not only Church's thesis and Turing's theorem, but all of recursive function theory; and we have on the other hand, a magnificent set of electronic devices that implement the mathematical principles; but as far as I can tell, no one has done the hard philosophical work of connecting the mathematics to the electronics in an intellectually precise way. What exactly is a computer? What is computation? What is a symbol? What is a program? We all know that notions like cognition and information and intentionality are not very precisely defined but we have been inclined to think - or at least I was inclined to think - that since these other notions sound so technical, they must be precisely defined. Your university and mine make tenure decisions on the assumption that we know what these expressions mean, so there ought to be a precise definition.

Well, someone might say, we don't have a precise definition of notions like "chair" or "table" or "bathtub" either. We don't have a philosophically precise definition of these notions but all the same, furniture makers and bathtub makers manufacture objects that we use successfully; so why should we be worried about precise definitions? Similarly, it seems to me that makers of actual computers that you can buy in the store don't have to worry about the definition of computation, they just have to get something that will work and will implement your wordprocessing and other programs. But if somebody says that contrary to physical appearances, the brain is a digital computer implementing the formal symbols in the programs which constitute the mind, then the absence of precise definitions does become worrisome, because without some such explanation we do not know exactly what is being claimed about the brain. What exactly is meant by these claims? So, let us go back and read Alan Turing's original definition. As I said earlier, he tells us that a Turing Machine is a device that can perform four and exactly four operations: it can erase a one and print a zero, it can erase a zero and print a one. It can move one square to the left, it can move one square to the right. It has a program of instructions of the form, if C then A, if condition C is satisfied, then perform act A, and it carries out these instructions very rapidly by printing and erasing symbols on a tape.

Now, I am a very literal minded person and on reading Turing, my inclination is to think that if I opened my computer, I ought to be able to find a lot of zeros and ones or at least a tape. But in fact, if I open the machinery, I won't find any such thing. Well, it doesn't matter, the theorists tell us, because what we need to find is not an actual zero or a one, or an actual tape, but rather something to which we can *assign the interpretation* of a zero or a one, and something which performs the function of storing information that is performed by a tape. There is a very large, indeed an indefinitely large, number of ways in which these functions can be performed. But at this point, a crucial question presents itself to us: What fact about the system constitutes its being a digital computer in the sense of implementing a program? Let us now turn to that question.

To explore this question I need to introduce a fundamental distinction. I think this distinction is more basic to our metaphysical view of reality than such familiar distinctions as that between mind and body, or nature and culture, it's this: there is a basic distinction in the ways in which we categorize reality between those features of the world that are identified solely in virtue of their intrinsic properties and those that are identified relative to the interests of human users and observers. To take an obvious example, that something has a certain shape and is composed of a certain kind of material is an intrinsic feature of that object. But that the same object is a chair or a bathtub is not an intrinsic feature of the object. It is what I will call 'an observer relative feature'. The distinction, in short, is between intrinsic features of reality, such as mass, gravitational attraction, atomic structure, and photosynthesis, on the one hand; and observer-relative features of reality, such as being a chair, being a bathtub, or being a nice day for a picnic, on the other. In general, observer relative features of the world also have intrinsic features, but the identification of the object with an observer-relative feature specifies more than just an intrinsic structure, it specifies how that structure relates to our interest as users and observers. Thus, for example, every bathtub has to have a certain kind of structure or family of structures, in order to function as a bathtub, but being a bathtub is something that is an addition to these structures. Calling something a "bathtub" identifies a structure in virtue of how we choose to use it or interpret it. So, the distinction between intrinsic features and observer-relative features is a distinction not between objective and subjective features, but between those features of reality that exist quite independently of our interests and those features which exist relative to our interests. That the object in my wallet is composed of cellulose fibers and has chemical stains are intrinsic features, that the same object is a twenty dollar bill is an observer-relative feature.

Now we come to the obvious question - what about computation? Is computation intrinsic or is it observer-relative? Well, notice that it certainly is not arbitrary whether or not something is a computer. Only certain sorts of devices can function in actual practice as computers. But computation as such is observer-relative. *We do not discover symbols or syntax in nature. Something* is a *symbol* or a *syntax only relative to* an *assignment of a symbolic or syntactical or computational interpretation.* So, the reason we have so much trouble answering the question, "What actual fact makes it the case that such and such a computation is being carried out?" is that there is no intrinsic fact that is constitutive of computation. Rather, computation is assigned relative to our interests as observers and users of the system. We could not *discover* computation going on in nature independently of us, because computation is defined in terms of symbol manipulation and there are no intrinsic symbols in nature. Computation is assigned to, for example, electronic processes in a commercial computer because we have designed it and built it precisely so that we could use it to carry out computations. That the system has certain voltage levels, and certain state transitions between voltage levels is intrinsic to the system. That these voltage levels are regarded as symbols, and the state transitions as computational processes is relative to us as observers, designers, programmers, and users.

The upshot of our discussion is that Cognitivism is incoherent. The question asked by Cognitivists is, "Is the brain a digital computer?" And Cognitivism gives an affirmative answer. But if the question is, "Is the brain *intrinsically* a digital computer?" then the answer is, "*Nothing* is intrinsically a digital computer.". But if the question is, "Can the brain be assigned a computational interpretation?" then the answer is "A computational interpretation can be assigned to anything.".

You could not discover that the brain or anything else was intrinsically a digital computer because something is a digital computer, and something is computation, only relative to the assignment of a computational interpretation. If I add two plus two to get four, that process is intrinsically computational because I am consciously and intentionally making that very computation. But if I have a computer such that when I type in "2+2" it types out "4", the computer intrinsically does no computing at all. I am using the computer to compute with. To put this point with brutal simplicity, computers do not compute intrinsically, but only relative to users, designers, programmers, etc. A computer that you buy in the store is an electronic circuit, intrinsically an electronic circuit, and you as an intentionalistic agent can use it to compute with.

This argument is different from the Chinese Room Argument, but in the end it is much more devasting to the computational theory of the mind. The Chinese Room Argument showed that semantics is not intrinsic to syntax. What this argument shows is that syntax is not intrinsic to physics. The reason that multiple realizability is a feature of computational systems - the reason that there is an indefinite range of physical systems that can carry out any computation - is that computation is not intrinsic to physics anyway. Any system whatever, to which you can assign a computational interpretation, is a digital computer.

So far we have arrived at the result that every computation requires an agent who assigns a computational interpretation to the process. This renders Cognitivism incoherent. Is there any way out of this? Well, if you look closely at the standard cognitivist literature, it seems to me that there is a tacit way out of this, but it is a way that is unsatisfactory. The literature standardly employs a homunculus vocabulary. There is, at least on the surface, a homunculus fallacy. Thus, for example, in the very statement of Marr's thesis of the problem of vision, we require a homunculus to make the statement intelligible. The statement, you will recall, is that the problem of vision is to explain how the system gets from a two dimensional visual array to a three dimensional description of the external computationally world. But the whole notion of a *description* requires that somebody is reading the description. This is a very crude version of the homunuculus fallacy, but there are similar homunculus fallacies throughout the cognitivist literature. Chomsky's discussion of the rules of syntax suggests that somebody or something is applying the rules in a recursive fashion to get an output of syntactically well-formed sentences. And it is hard to see how to interpret that without thinking of somebody acting as a homunculus and applying the rules.

The homunculus talk which is standard in the cognitive science literature does not worry the partisans of Cognitivism because they think the homunculus is eliminable. In fact, there is supposed to be a generally accepted procedure for eliminating the homunculus: the homunculus is eliminated by replacing him with progressively stupider homunculi until eventually we reach a level of homunculi who can only perform simple binary operations. This elimination of intelligent homunculi by stupid homunculi is called "recursive decomposition". Here is how recursive decomposition is supposed to work: suppose you have a pocket calculator that multiplies seven times eight to get fifty six, and you ask, how does it do it? The first answer might be that there is an intelligent homunculus who knows how to do multiplication. But if you ask, "How does he do it?" he is replaced by a dumber homunculus who does not know how to do multiplication but who knows how to do addition and who takes eight and adds eight to it incrementally six times to get fifty six. Now, if we ask the homunculus who does the addition, "How do you do it?", this less intelligent homunculus is replaced by even stupider homunculi. He is replaced by a bunch of homunculi who say "We don't know multiplication, and we don't know addition, but we do know how to convert decimal symbols to binary symbols, and that's what we do." And if we ask them how they do that and what happens when they do it, what we discover is that they are replaced by an even stupider bunch of homunculi who say, "Well, actually we don't know about decimal and binary, and we don't know about addition and we don't know about multiplication, but we just apply simple algorithms to zeros and ones - that's all we do.". We keep asking this question "How do you do it?" and we keep replacing smarter homunculi with stupider homunculi until finally we reach the level of the stupidest homunculi of all, and they simply say "zero, one, yes, no, flip, flop.". In short, the homunculus talk at the higher levels where we had to postulate intelligent homunculi is just a manner of speaking. There weren't any homunculi at those levels. They are replaced by progressively stupider homunculi until finally we 'discharge' the intelligent homunculi with a whole army of extremely stupid homunculi who just perform a simple binary operation.

Does this recursive decomposition remove the homunculus fallacy? It plainly does not. It was designed to answer a different question. It is designed to answer the question, "How can more intelligent processes be composed of less intelligent processes?". But both the more intelligent processes and the less intelligent processes require a syntactical attribution to the physical processes in question. Thus, we still require a homunculus and Cognitivism does not avoid the homunculus fallacy through recursive decomposition. The upshot is that Cognitivism remains incoherent because the thesis that the brain is a digital computer is ill-defined. If the claim is that the brain is intrinsically a digital computer, we know the claim is false, because nothing is intrinsically a digital computer. If the claim is that a computational interpretation can be assigned to the brain, we know the claim is vacuous, because a computational interpretation can be assigned to anything at all.

IV.

I said that our age, like all the others, has its own forms of absurdity. Our grandchildren will wonder how we could have believed anything so silly as the computational theory of the mind. But our absurdities, like earlier ones, come from deep confusions and satisfy deep needs. The computational theory of the mind is one of the final expressions of a confusion that is deep in our culture: Mind-body dualism, the belief that there is a distinction between the mind and physical reality. And computationalism satisfies a deep need: on the computationalist view, the mind is not part of ordinary biological reality, like digestion or excretion. No. On the computationalist view, the mind is something formal and abstract, a beautiful and ethereal set of computational operations. And therefore computationalism is perhaps the purest expression of the Cartesianist need to deny the brute physical, biological mode of our real existence. Only when we overcome this confusion and see that this need cannot be satisfied will we fully overcome this and related forms of contemporary absurdity.

Prof. Dr. John R. Searle
Department of Philosophy
University of California
Berkeley, California, 94720, USA

NOTES

1 This paper was originally given as an informal talk, and I have left the conversational style of the original presentation in this published version. The ideas are developed at greater length and with more supporting argument in my book, *The Rediscovery of the Mind*, MIT Press, Cambridge 1992.

2 San Francisco, W.H. Freeman and Co., 1982.

3 *The Computer and the Mind*, Cambridge, Harvard University Press 1988.

4 *Computation and Cognition*, Toward a Foundation for Cognitive Science, Cambridge, MIT Press 1984.

5 "The Computer Model of the Mind", in: D. Osherson and E.E. Smith (eds.), *An Invitation to Cognitive Science 3*, Cambridge, MIT Press 1990, pp. 247-289.

PART III

TOWARDS A SCIENCE OF MAN AND SOCIETY: PHILOSOPHICAL AND METHODOLOGICAL FOUNDATIONS

SELF-ORGANIZATION: NEW FOUNDATIONS TOWARDS A "GENERAL THEORY OF REALITY"

JOHANN GÖTSCHL

SUMMARY

In this paper, an attempt will be made to show that the fundamental conceptual categories of the Theory of Self-Organization (TSO) furnish novel opportunities for depicting man's position in the world. Preparatory work towards a theory of process-evolutive understanding of reality where the structure and dynamic of nature and culture can be represented in a unified way becomes possible. Then we can identify a convergence between the understanding of reality as mediated by the "Science of Nature" on the one hand, and the understanding of reality as propagated by the "Science of Man" (self-understanding) on the other. The philosophical reconstruction of the meaning of TSO leads to the realization that a new type of science can be developed; it will be a symbiosis of the "Science of Nature" and the "Science of Man". Up to now, the "natural sciences" have not been able to provide a suitable approach to man or society as a whole. A similar situation, but linked to much more serious problems, can be found in the humanities and the social sciences; both of them have so far also failed to provide an adequate approach to nature as a whole. TSO fundamentally changes the situation: it is an increasingly integrated component of thinking in the categories of the natural sciences; and this makes possible a new type of modeling social processes. As a naturalistic system of science, TSO does away with the apparent difference between the categories of material and social structures. The implementation of TSO in the categories of social-scientific thinking shows that, for the first time, a socio-scientific approach to nature has become possible. Any efforts in the direction of identification and development of new presuppositions for establishing a "general theory of reality" will share an ultimate goal: to grasp the human thinking process itself as reality; and this means, finally, to show that the socio-cognitive structures can be imbedded in reality in a consistent manner.

(I) CONVERGENCE BETWEEN "NATURAL SCIENCE" AND "SOCIAL SCIENCE"

New knowledge derived from the theory of self-organization (TSO) of specific real systems constitutes a new understanding of nature and culture, or society; it implies a new self-understanding, too. In its core, TSO contains the principles of emergence, self-production and creation of new forms or

J. Götschl (ed.), Revolutionary Changes in Understanding Man and Society, 109–128.

structures; in this way, TSO leads to the possibility of a categorially new, highly differentiated process-evolutive concept of reality, and thus it breaks up the traditional scientific classifications and the different kinds of scientific ontologies. TSO imbues the notional categories of configuration and transformation of material systems with new significance. This has a reason: according to TSO, the principles that lead to the emergence of dissipative structures are the very same ones which, under specific conditions, lead to a separate, self-organizing dynamic in material systems. This separate dynamic is described by the principles of energetic and informational openness and of operational closeness as well as by the conditions for the instability and stability of systems. These principles of a separate dynamic concern material systems; they serve as the foundations for a more unified and process-evolutive conceptualization of reality, a reality which no longer solely refers, or is restricted to, material systems.

This process-evolutive concept of reality attaches itself no longer to the configurations of the inorganic, organic or social systems which require categorial differentiation. Contrary to what has been understood by this up to now, it is here intended to mean the interwoven functions of 1) understanding, 2) explanation, 3) predictability and 4) realization. TSO leads us to a model of reality in which the framework for the compatibility of determinism with specific unpredictable developments in material systems is based on the concept of bifurcation, i.e., the concept of the unity of linearity and non-linearity in the evolution of systems. The compatibility of determinism and uncertainty and the linking of linearity and non-linearity have thus become the new prerequisites for the construction of a universal, process-evolutive understanding of reality.

However, a generalization or synthesis of the classical theories of the natural and social sciences which could bridge the apparent categorial gap between material and social systems is still not possible. This is so because the subject matters which are fundamentally posited as ontological correlates in the "natural sciences" and as unalterable prerequisites in the (traditional) "human sciences" are not represented through self-organizing characteristics. However, new breakthroughs are being made with TSO; new prerequisites for a more unified description of the multiplicity of reality are coming into being; TSO represents a new unity in reality's multiplicity.

Particularly since Newton, the sciences have changed rapidly. New theories have emerged; but methodology and ontology have remained more or less the same. TSO drastically changes the situation: theories, methodology and ontology are appearing in a new, interconnected way. With the help of the philosophical reconstruction of TSO the ontological correlate of classical natural science is gradually being recognized more accurately. This correlate of the classical physical theory (CPT) is a substance-based ontology which determines the subject matter. The five familiar principles of classical physical theory (CPT), summed up here in philosophical reconstruction, are: (1) the principle of the invariance of mass; (2) the principle of linear causality; (3) the principle of the limited potentiality of matter; (4) the principle of reversibility (principle of symmetry); and (5) the principle of the categorial separateness of mind and matter, object and subject (principle of non-coherence of object and subject).

These five well-known principles mirror what can be called substance ontology. On the physical level of terms, this paradigm corresponds to Newton's concept of reality, the two central characteristics being those of (i) mass, which is understood to be constant (invariant); and of (ii) movement which describes dynamic segments of reality but cannot explain the origination, stability or instability of new forms and structures. It falls short of supplying exactly those scientific characteristics for a process-evolutive concept which are, in fact, provided by TSO.

Although the theory of relativity and quantum theory differ in many respects as far as the conceptualization of reality is concerned, they both contribute something new: substance vanishes almost completely, with the result that all five characteristics of classic substance ontology are specifically suspended to a degree. Essentially, this loss of substance takes place in the conceptual framework of the theory of relativity by means of the matter-energy equivalence relations, together with the field-theory representation of reality, and in the conceptual framework of quantum theory by means of transforming the notional object, together with the interdependence of subject-object relations. For this reason, it is helpful to speak here of functional ontology. This functional ontology is, in fact, a far-reaching process concept of reality. But here, too, the relevant characteristics of evolutive processes of self-organization are missing.

Using TSO, this paper's central hypothesis, then, can be formulated in three versions which become increasingly more concise. The second version, for instance, allows for a tendency towards suspension of the apparent categorial difference between material and social systems. The third version of the central hypothesis refers to a tendency towards transcending the mind-matter duality hypothesis.

Central hypothesis (1st version):

TSO results in a radical improvement: a **more unified conceptual representation** of reality. Material systems and social systems can increasingly be represented in a more unified way. This **more unified conceptual representation** of two apparently categorially different realities analyzes analogies more deeply; the result is that these analogies are eliminated through a process-evolutive, universalizing modeling of reality.

TSO allows the identification and an increasingly differentiated explanation of presuppositions necessary for a new classification of science. This has radical consequences for the construction of a "general theory of reality". What are these prerequisites? The central point is that it is not claimed that TSO has found new characteristics of material systems which can simply be added to the characteristics which material systems have so far been stated to have. The generating principles state that reality must be modeled along a dominant configuration and transformation of elements and structures with its own dynamic. There are characteristics that lead to the generation of coherent structures or shapes, respectively, in material systems, and which are, within certain limits, an immanent part of the system. The notion of material systems (or subject type) constituted by the principles of TSO represents material systems in a way where areas of reality which have up to now appeared to differ categorially, e.g., material processes and social processes, can be increasingly bridged theoretically in a process-evolutive way and without reductionism.

The development of TSO permits us to differentiate more clearly between scientifically constituted conceptions of reality and their corresponding ontologies. The different kinds of ontology constituted by science can be separated from one another as in the following minimal classification:

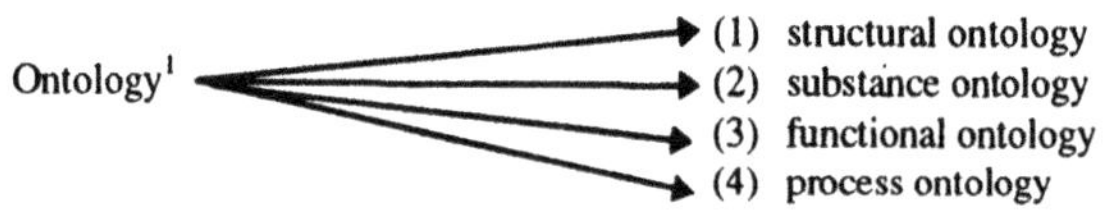

A brief interpretation of these four predominant, scientifically constituted types of ontology is indicated by 1) - 4):

1) is essentially constituted by mathematics, logic and pure system theory;
2) is essentially constituted by classical physical theory (CPT);
3) is essentially constituted by the theory of relativity (RT) and quantum theory (QT);
4) is essentially constituted by the theory of self-organization (TSO).

What results from the above in terms of an identification of relations between "natural science" and "social science" in general, and in terms of a proof of a convergence between the two in particular? From an epistemological standpoint, three insights are of importance. These are: (i) TSO functions as a bridging science; (ii) TSO functions as a science of integration; and (iii) TSO functions as a universalizing science. What are the most important characteristics?

In the course of the 20th century, all reductionist attempts to create a more unified model of reality, of material and social processes have failed. This failure is mainly due to the fact that reduction in the sciences and philosophical reductionism have not been sufficiently differentiated. Reduction is indeed indispensable; but reductionism is not. The attempt to trace all phenomena back to logical/mathematical and purely system-theoretical structures has failed. Even the less ambitious attempt to trace everything back to physicalist structures has failed. Essentially, there are three reasons for this unsuccessful outcome. TSO, however, opens new opportunities; they do not solve the problem of reduction altogether but bypass it, so to speak. The failure of this reductionist program can be explained by three reasons:

1) It is not possible to provide a satisfactory empirical interpretation of logic, mathematics and pure system theory.
2) If we attempt to unify the notional representations or even to homogenize the material and social structures, the complexity rises to an unmanageable degree.
3) The dynamics of material and social structures are too great, given certain conditions.

TSO has brought about fundamental changes for all three structural features by supplying the basis for this change: 1) can be bypassed by using the method of mathematical computer simulation (taking into account fractals). 2) and 3) can be bypassed by using the central element of TSO, namely, the concept of dissipative structures. Increasingly, TSO is being incorporated as an immanent component into mathematical computer simulation. This makes it possible to derive more unified terms from the complexity and dynamic of material and social structures. The linking of mathematics and computers which results in a new type of processing in the form of computer simulation represents a new methodological element. In computer simulations, material, notional and visual dimensions of thinking about reality interweave, and they interweave in reality. At the same time, TSO is the theory of complex and dynamic systems; gradually it identifies those characteristics for process-evolutive material systems which show a strong affinity to the characteristics displayed by social systems. For the first time in the history of science, this has produced better reasons for examining social structures by means of conceptual frameworks of material structures and, vice versa, for examining material structures by means of social structures. This possibility - although it still needs to be worked out in detail - is leading to a new understanding of the interfaces between "material structures" on the one hand and "social structures" on the other. Put more strongly, the following supposition can now be better justified:

Central hypothesis (2nd version):
TSO supplies two new results, particularly in the context of computer simulation support: (i) a tendency towards ontological neutralization and, simultaneously, (ii) empirical adequacy which makes possible a more unified representation of material and social structures.

These two characteristics - far-reaching ontological neutrality and, at the same time, empirical adequacy - make it possible to proceed from mere analogies between "material structures" on the one hand and "social structures" on the other to evolutionary process models. In turn, these models make possible a more unified notional representation of material and social phenomena. This scientific advance to evolutionary process models is supported by the following familiar characteristics of systems which have been identified by TSO. It is mainly these characteristics which qualify TSO to manage without recourse to physicalism and biologism.

TSO demonstrates that:

1) All real systems are open systems. They are not in an equilibrium: processes are dissipative, chaotic and self-organizing. As sequences of dynamic patterns they are of a process-evolutive character.
2) Under certain circumstances, systems can be self-determining, i.e., they influence themselves. This can lead to levels of configurations which cannot be reduced to other levels. There are limitations for system reductions.
3) In dissipative, self-organizing and irreducible processes, there is no necessity to postulate the general existence of hierarchic principles of formation and development.
4) Systems can have the structure of self-referentiality and recursivity; i.e., in case a system unfolds in a process-evolutive manner, the dominance of the internal coherence of a system reduces and relativizes the meaning of factors external to it.
5) System-immanent characteristics of co-operation and co-evolution lead to a further differentiation of the identity (self-identity) or the autonomy of the system.

On the basis of 1)-5), a decisive characteristic can be formulated which enables us to establish a better and unified process-evolutive modeling of material and social systems.

6) TSO provides a new stipulative approach to the relationship between macro and micro levels by using circular determinations. This means

that the elements of a system cause the regularities (patterns, order); and the regularities cause the order of the elements. Therefore, it is rational to assume that system elements have a certain stability.

On the basis of 1)-5), 6) shows that TSO gradually establishes a notional representation of physical and social systems in a more unified way. This leads to the elimination of the classic problem of reduction, especially in view of the fact that there is no reduction of classical physics, RT or QT, or the theory of evolution, TSO. Neither is there a reduction in the other direction; rather, TSO identifies other characteristics of systems which give the terms "object" and "reality" a new meaning. That is to say that the ontological correlate of TSO is a process-evolutive concept of reality or TSO makes two things possible: (i) a more unified foundation of the natural sciences (physics, biology); this is already well developed. (ii) Beyond (i), TSO yields a more unified foundation of our understanding of material and social systems.

Put in a diagram:

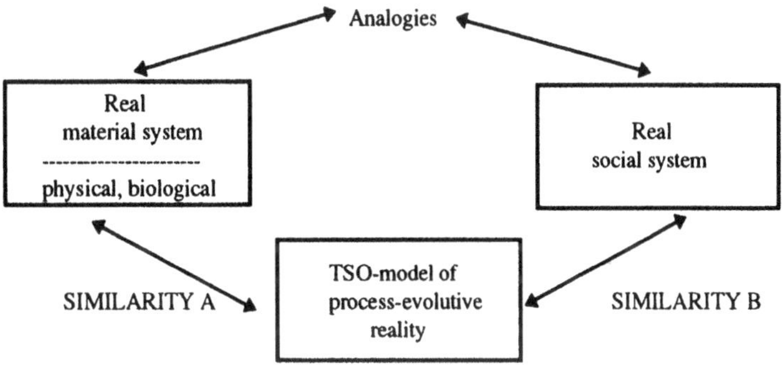

The philosophical reconstruction of TSO validates:

Similarity A $\approx$ Similarity B.

This means, a reduction is based on mere analogies or, put differently: the construction of "process-evolutive models of reality" tends to substitute for the use of analogies. Here the outline of a paradigmatic change emerges.

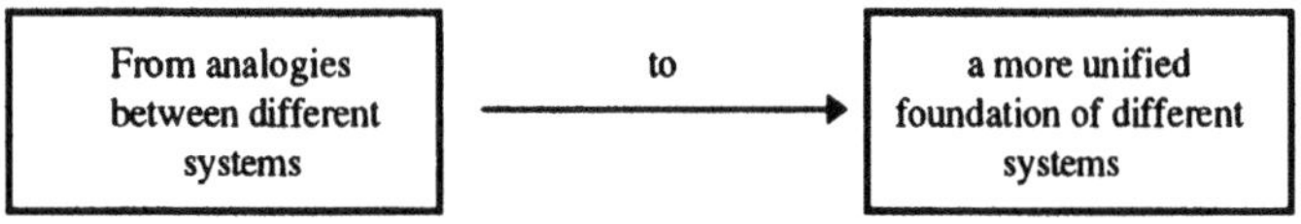

The approach which establishes that similarity A ≈ similarity B has been well documented over the last few years. TSO has shown the general interface of cosmic, biotic-ecological and cultural (social, economic) self-organization. As is well-known, the central point here is the inner context between irreversibility (arrow of time) and self-organization (negative entropy as creator of order).

The philosophical reconstruction of TSO's area of validity shows, however, that one important limitation caused by an as yet unsolved problem needs to be mentioned: according to QT, a physical system must be understood as the totality of material processes with specific interconnections between micro and macro levels. Through QT, TSO appears as a specific macro theory. The validity of TSO for objects and processes belonging to the quantum world has not yet been proved, although currently there are no good reasons to doubt TSO's representation of the quantum world. Working through this aspect reveals that the similarities between physical and social systems, the reduction implied by these similarities, mere analogies between quantum-theoretical representations of physical systems and the representations of social systems by TSO might reach a limit. As far as the QT representation of material systems is concerned, one cannot as yet claim that a tendency towards a unified representation of physical and social reality does exist. This means that similarity A is, for the time being, limited to the molecular dimension. Since it is both a macro and a micro theory, TSO cannot currently represent the QT dimension; from this it follows that certain implied assumptions as to a unified representation of nature, man and society must be made for heuristic reasons. The most important and highly plausible implicit assumption is: at the moment, the theoretical chances of developing QT model-theoretically in such a manner that social processes could be represented by it appear to be low. Much higher are the chances of developing TSO model-theoretically so as to represent the QT dimensions of material reality.

(II) COGNITION, COMPLEXITY AND SELF-ORGANIZATION

The simultaneous increase in ontological neutrality and empirical adequacy leads to a more unified conceptual basis for the construction of a "general theory of reality". This reveals that TSO brings with it a new structuring for the existing canon of science. In its main features, this new order consists in accounting for the already mentioned three-fold functionality:

1) TSO as a bridging science;
2) TSO as an integrating science;

3) TSO as a universalizing science: it establishes complexity, constructs convergence between the individual sciences and stipulates new relations between inner and outer world by means of linking the individual sciences.

TSO's complexity, then, affords the transition from substance and functional ontology to a process-evolutive onotology. Here particularly the third function plays an important role: it leads to a new definition of the relationship between evolution and cognition. This new definition allows to formulate the third and more concise version of the central hypothesis. It will be shown how this more concise hypothesis is a viable alternative to the dualistic conception of reality.

Central hypothesis (3rd version):
As a theory of the complexity and dynamics of reality, TSO identifies and represents the existence of new affinities between evolution, self-organization and cognition. These affinities lead to progress in overcoming the dualistic mind-matter hypothesis.

Epistemologically speaking, this hypothesis can be expanded: with the help of TSO one realizes, that, from a certain notional level of development, the scientific knowledge of the outer world (knowledge of objects) represents the potential knowledge of the inner world (self-knowledge). A process-ontological understanding of the world, of nature and of matter mediates in a non-reductionist way between the various levels of structure (material and social ones) by means of a continuously improving knowledge of the conditions for the occurrence of various structural levels. At this point, one discerns that the classical canon of scientific classifications according to which, for ontological reasons, a differentiation has to be made between natural and social sciences can no longer be maintained. Although the ground for knowledge of the unity of natural and cultural history has been prepared even before the time of TSO, it has only been through TSO that considerable progress in this direction has been made. Due to TSO, the coupling processes between material and social dynamic can be identified and explained. The classical differentiation of the sciences according to "given" objects and a "given" ontology has obviously been left behind. The "sciences of nature" and the "sciences of man" constitute a synthesis, even though we have as yet no name for it. Let me call it "general theory of reality". A recourse to the philosophically reconstructed TSO shows that the classic differentiation between the natural and the social sciences is about to expire.

The central hypothesis' third version, in particular, is supported by three characteristics of TSO which indicate the way to overcome the dualistic model of reality.

Characteristic A:
TSO functions as a theory of complexity.
Characteristic B:
TSO functions as a theory of how sciences converge.
Characteristic C:
TSO functions as a theory of how external reference (natural sciences) and self-reference (psycho-social sciences) interweave.

These three functions - complexity, convergence-forming and interweaving - tend to replace the dualistic concept of reality with a process-evolutive monistic model of it. It is evident that the ontology of substance fails completely; and that the ontology of function fails in part when it comes to the central category. What they both lack are modal categories for a theoretical modeling of realities in terms of a process-evolutive system on various levels of structure. This deficit will only be corrected by TSO because TSO shows the possibilities and limits within which systems organize themselves and emerge. Due to its specific scientific approach, TSO conquers, for the first time, the dualistic mind-matter concept. A brief analysis of the three characteristics should help to explore the limits and possibilities of conquering the dualistic mind-matter concept.

Re A) TSO as a theory of complexity:

Recent scientific developments point to a new tendency: more and more scientific results function as scientific bridges which lead to a weakening of the mind-matter dualism. There are mainly two tendencies which were first developed independently of TSO; but TSO expands them theoretically. Let me make two brief preliminary remarks about this:

Preliminary remark (1): The rapid elaboration of the interfaces between genotype and phenotype led to the recognition that human consciousness can only be modeled, in an empirically adequate manner, as a specific emergent configuration or structure of neural processes. The whole spectrum of biological sciences which deal with the apprehending subject - from macro-biology to the most highly differentiated molecular biology, neuro-biology, neuro-physics, neuro-chemistry, neuro-linguistics and neuro-psychology - reveals itself as an integrative, universal imbedding of the subject in the context of nature. This integration transfers both the natural term and the term of social or cognitive systems to the categorial representation, interpretation and modeling frameworks of a naturalistic, complex system term.

Preliminary remark (2): The modeling of nature in the categorial framework of terms of complex naturalistic systems leads to new possibilities; by now, they have become well-known. Generally speaking, this concerns computer simulation on the basis of material and social (symbolic) systems: mathematics, informatics and technology. These have made it evident that a technology (computer) constructed on the basis of natural laws can represent system processes which grant insight into its universalizable degrees of freedom. This, too, is a new form of inserting the apprehending subject (the "self") into a complex naturalistic system term.

These two preliminary remarks explain the assumption that it is the complexity of TSO which leads to progress in the attempt to eliminate the dualistic hypothesis of reality. In brief: the level of complexity of TSO can be seen from the following characteristics; they are of central importance for a philosophical analysis if we want to overcome the hypothesis of duality.

A philosophical reconstruction and expansion of TSO has led to a new, more general term of object or reality because the ontological correlate of TSO consists in a process-evolutive conception of object and reality. When we compare the use of the term "reality" in TSO with its use in classical physics (CPT), Quantum Theory (QT) and the theory of relativity (RT), we get:

reality concept (CPT) ≠ reality concept (QT, RT) ≠ reality concept (TSO).[2]

TSO constitutes a reality term which leads to a series of epistemological or ontological implications. This is tied to the notion that TSO indicates a unity "behind" the multiplicity of material and social systems. This unity in multiplicity increases the representational and explanatory values of concepts according to which the differentiated and interwoven structural levels of biological and social complexities have evolved from material structural levels. Essentially, this means that TSO categorially expands our knowledge achieved up to now, without, however, being incompatible with RT and QT. The essence of TSO is that complex systems can increase in degrees of freedom by themselves. This recognition is a categorial new step in the development of science, since, as has been shown earlier, TSO - in contrast to, and as expansion of, RT and QT - allows for a decisive, twofold reference which is due to its ontological neutrality and empirical adequacy:

i. a reference to material structural levels;
ii. a reference to cognitive structural levels.

Expressed diagrammatically:

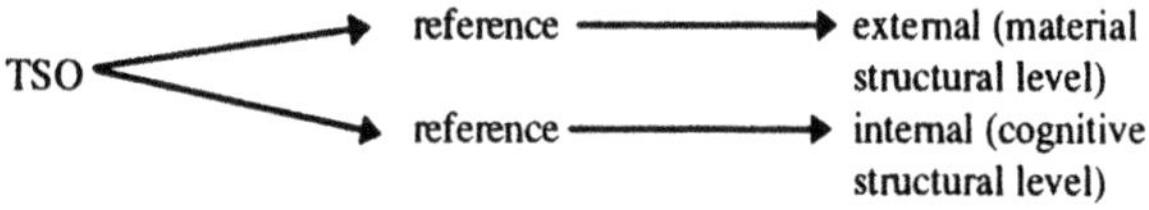

This twofold reference structure of TSO can be given the following philosophical interpretation: with the help of the complex descriptive and explanatory potential of TSO, we can express the knowledge that the referential objects of TSO are not only the objects which are opposite the apprehending subject but, to a certain degree, also the relations between subject and object. With the emergence of TSO, then, a structural interwovenness of object, knowledge of object and self-knowledge can be identified. Against the background of a twofold reference potential of TSO, the following, more complex approach can be taken as a starting point for a conceptual unfolding of the central hypothesis (third version!):

Complex approach: a scientific theory (Ti) as conceptual framework shall be categorized as more complex than a scientific theory (Tj) if (Ti) can represent more differentiated elements concerning co-operative and evolutive structures as ordered totality than (Tj).

This complex approach, a necessary but not sufficient criterion, can, by means of a diagram, indicate the new principles for the construction of a "general theory of reality".

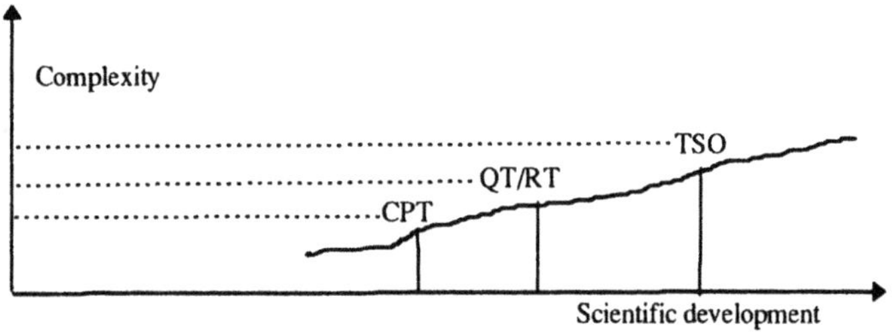

This complex[3] approach reveals that, in scientific development, it took TSO to advance to a more unified theoretical-notional representation of material and social (cognitive) structural levels. Thus, complexity (Comp) in the sense of the above complex[4] approach enters into the following formulation:

$$\text{Comp (CPT)} < \text{Comp (QT, RT)} < \text{Comp (TSO)}.$$

This result is founded mainly on the principles of a very strict ontological neutrality and empirical adequacy on the one hand, and the - closely connected - principle of a twofold reference potential on the other. It permits implications which signal familiar transitions. These transitions form the categories for the construction of a "general theory of reality" as a symbiosis of material and socio-cultural structural levels. The main transitions are:

- Naturalistic understanding of a system instead of idealism;
- Monism instead of idealism;
- Holism instead of particularism;
- Emergentism instead of (classical) determinism;
- Self-organization instead of exogenous organization.

All this leads to an increasingly better justification of the central hypothesis. As a naturalistic theory of complexity, TSO offers the possibility of a more comprehensive, naturalistic-systematic self-interpretation of man. The unification of the dimensions of nature, man and society becomes a scientific option.

Re B) TSO as a naturalistic system science:

The twofold reference structure of TSO identified by means of philosophical reconstruction can be made more explicit by two further characteristics. So far, analyses have come up with possibilities which establish the following: the reference potential of TSO is such that there is the possibility of a more unified notional representation of material and social (cognitive) structural levels:

In order to make this aspect clear, two levels of analysis can be used. One level concerns the specific interfaces between socio-cognitive and physico-material structural levels. The other level of analysis refers to a brief comparison of dominant indicators.

Re analysis level 1. One of the classical methodological and ontological models of cognition prescribes that subject and object have to be separated if we want to understand the external (material) world. To understand the external (material) world means to produce symbolic and experimental interactions and aggregations with it. This leads to specific social and cognitive phenomena, such as identities, forms and totalities. An ontological assumption, which receives its more precise meaning from TSO, is at hand: there is a definitive correlation between the state of the socio-cognitive world on the one hand, and the physico-material world on the other. "Understanding", for instance, represents the state of correspondence between the socio-cognitive and the physico-material system. On the basis of TSO, understanding, then, obviously represents the existence of a dynamic equilibrium of the cognitive system and, simultaneously, the existence of a dynamic dis-equilibrium of the neural material system. These interfaces will require further theoretical exploration; but they already suggest that, and how, socio-cognitive structural levels are linked with material (neural) structural levels; they also indicate through which parameters the socio-cognitive and the material structural levels can be identified as interdependent totalities.

The twofold reference potential of TSO, namely, the reference to material and to social (socio-cognitive) structural levels, changes the meaning of self-understanding, of "self", for "self" becomes evident as a specific totality of material and socio-cognitive structural levels if we categorize in TSO:[5]

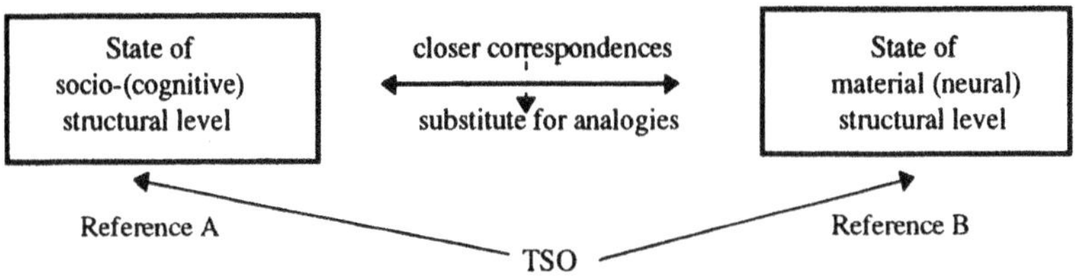

Reference type A (internal reference of TSO) and reference type B (external reference of TSO) make it possible to say that the fundamentals are about to get one step closer towards a "general theory of reality". As a naturalistic system science, TSO develops an approach which, up to now, has been unthinkable: the sociological or human-theoretical access to material structural levels, to the physical world and to nature.

Re analysis level 2. If we want to improve compatibility, we have to carry out a comparative analysis of scientific criteria. This will enhance ontological neutrality and empirical adequacy, as mediated through TSO's twofold reference structure.

So far, the analysis reveals that TSO supplies a new theoretical model of reality. What is central to this model is its process-evolutive character; this makes possible a more unified foundation of material and socio-cognitive structural levels. With the philosophically reconstructed TSO the preconditions for the construction of a "General theory of reality" become increasingly identifiable. This means that the theorems of philosophically reconstructed TSO which permit the foundation of a system of cognition mature: they evolve into a heuristic, fruitful core of scientific development for both the "natural" and the "social" sciences:

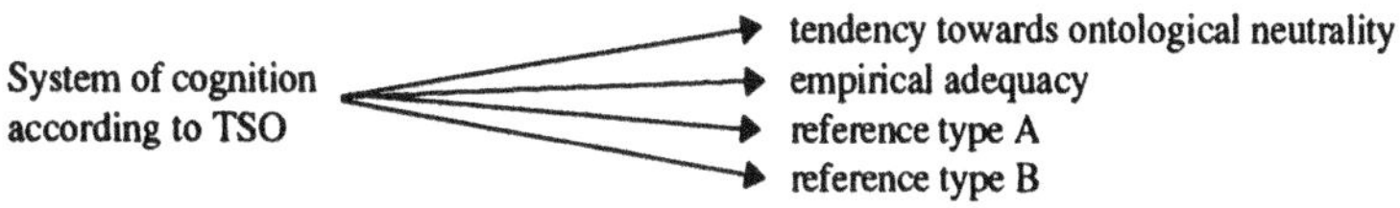

We see the outlines of a possible development: the theories of the philosophically reconstructed and philosophically expanded TSO concepts may evolve into an immanent component of the natural scientific approach in the dimension of a socio-cognitive access to natural phenomena; a socio-scientific access to natural phenomena on the other hand reduces the familiar problems of physicalism and biologism, analogism and projections.

Let us make a matrix for the central characteristics (criteria) which distinguish the types of science of the 20th century from one another, i.e:

Type A: NSC ❍ Natural sciences
Type B: SOSC ❍ Social sciences
Type C: TSO ❍ Sciences of self-organization.

Scientific criteria	NSC	SOSC	TSO
1) Clarity	yes	yes	yes
2) Consistency	yes	yes	yes
3) Reproducibility	yes	no	no
4) Predictability	yes	no	no
5) Objectivity	yes	no	yes
6) Reducibility	yes	no	no
7) Truth	yes	no	yes
8) Historicity	yes	yes	yes
9) Quantifiability	yes	no	yes
10) Sensitive dependence on initial conditions	no	yes	yes
11) Recursivity	no	yes	yes
12) Self-referentiality	no	yes	yes
13) Creativity	no	yes	yes
14) Knowledge of not-knowing	no	yes	yes
etc.			

When comparing NSC and SOSC, we find at least 11 incompatibilities among the 14 comparative indicators; this underlines the principal direction of development. When comparing SOSC and TSO, we find only three.

What results from this for the complex questions in connection with the new presuppositions which are relevant to the construction of a "general theory of reality"? The answer to this question by means of 2nd level analysis has been enriched by at least one aspect: the philosophical reconstruction of TSO shows that TSO as a naturalistic and systemic science on the one hand and as a social science on the other reveals interesting affinities (convergences).

Re C) TSO as a basis for identifying a linking rule:

The deeper exploration into the structural formation and structural dynamic of material and socio-cognitive structural levels leads to the insight that the level of scientific development which is represented by the categories of TSO culminates in one central aspect of knowledge: knowledge of the external material world has a specific meaning as knowledge of the internal world. In other words: on the categorial level of a philosophically reconstructed TSO it becomes apparent that each perception of the external world represents a potential perception of the internal world; as the nucleus of our perceptual production of the external and internal world, TSO permits the philosophical reconstruction of itself and scientific self-reference as well. Thus, it also becomes clearer that material and social structural levels, "mind" and "matter", are coherent constructs; i.e., foundations for the construction of a "general theory of reality" become evident.

It can be seen from two characteristics that this is a further decisive aspect:

i. More and more "new" sciences allocate themselves in between the classic "natural sciences" and "social sciences". The border areas between these two forms of science become more and more densely "populated".
ii. More and more sciences coalesce.

If we want to explain aspects (i) and (ii), this is easier when using the TSO system of cognition. The increasing "density" and linkage between material and socio-cognitive structural levels can be seen, among others, from the rapid developments in "more recent" sciences adduced in the following table. No claim is made as to the completeness of the list. Even this selection shows that the two extremes "mind" and "matter" are subject to paradigmatic change, irrespective of the duality hypothesis which posits a fundamental discontinuity. Risking an overstatement, the assumption could be made that the development of science will transform the classic discontinuity of "mind" and "matter" into a continuity:

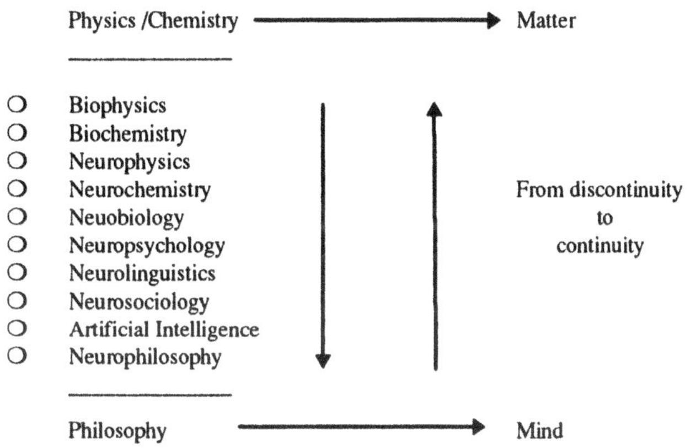

When we focus on the decisive aspect of development, we see that the tendency towards a continuum between mind and matter is relative to the coalescence of the sciences:

In a diagram, it looks like this:

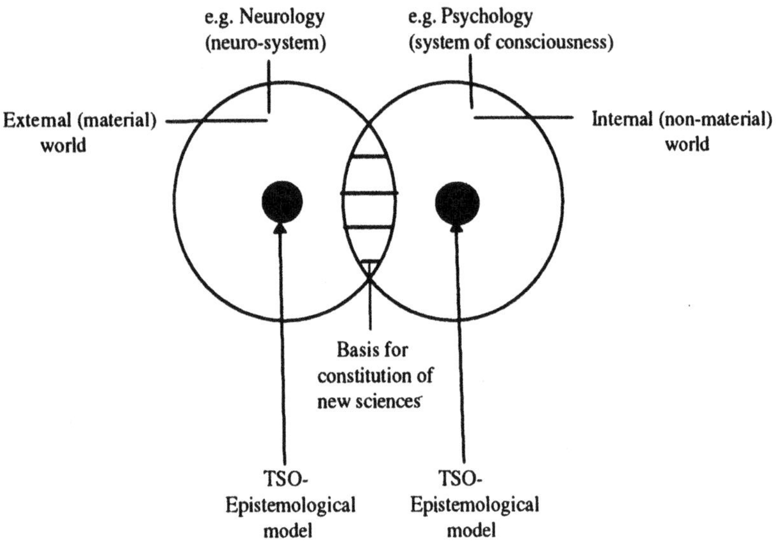

The diagram illustrates one important aspect: the process-evolutive modeling of reality by means of the TSO system of cognition facilitates the constitution and identification of links between individual sciences. This type of linking rule shows once more that the scientific conditions for establishing a "General theory of reality" improve. In such a theory, the apparently categorial differentiation between mind and matter will no longer be a scientific necessity.[6]

Conclusion 1:
This philosophical exploration of TSO has outlined the possibility of a "general theory of reality".[7] However, the ultimate goal of such a theory will not be achieved for some time, since it needs to be shown that the human thinking process is consistent as a real process and can be theoretically imbedded in reality.[8]

Conclusion 2:
This philosophical exploration of TSO - from which the TSO system of cognition can be developed - reasons that nature and society, mind and matter are coherent constructs.

Conclusion 3:
TSO argues that a higher ontological affinity, and probably even the same ontological status, will have to be allocated to man and nature.

Prof. Dr. Johann Götschl
Department of Philosophy,
Ludwig Boltzmann Institute
for Science of Research,
University of Graz
Mozartgasse 14, 8010 Graz, Austria

NOTES

1 To be understood as scientifically constituted models of reality. These models as different kinds of ontology are understood here to mean the result of a philosophical reconstruction of scientific theories. For more details compare Götschl, J., "Philosophical and scientific conceptions of nature and the place of responsibility", in: *International Journal of Science Education*, 1990, pp. 288-296 und Götschl J., "Hypercritical physical realism and the categorial changes in the subject-object relations", in: *La Nuova Critica*, 1991, pp. 5-19.

2 For details compare 1. Götschl, J., "Philosophical and scientific conceptions of nature and the place of responsibility", in: *La Nuova Critica*, 1988. 2. Götschl, J., "Hypercritical physical realism and the categorial changes in the subject-object relations", in: *La Nuova Critica*, 1991. 3. Götschl, J., "The role of physics in Erwin Schrödinger's philosophical system", in: *Erwin Schrödinger: Philosophy and the Birth of Quantum Mechanics*, M. Bitbol and O. Darrigol (eds.), Editions Frontieres, Paris, 1992. 4. Götschl, J. and Leinfellner, W., "Erwin Schrödinger's world view: The role of physics and biology in his philosophical system", in: *What is Controlling Life? 50 years after Erwin Schrödinger's What is Life?*, E. Gnaiger, F.N. Gellerich and M. Wyss (eds.), *Modern Trends in BioThermoKinetics*, Vol. 3, Innsbruck: University Press, 1994.

3 Götschl J., "Hypercritical physical realism and the categorial changes in the subject-object relations", in: *La Nuova Critica*, 1991, pp. 5-19.

4 Simon, H.A., "The architecture of complexity", in: *Proceedings of the American Philosophical Society 106*, 1962, pp. 467-482.

5 Götschl, J., "Physics in Erwin Schrödinger's philosophical system", in: *Philosophy and the Birth of Quantum Mechanics*, M. Bitbol and O. Darrigol (eds.), Editions Frontières, Paris, 1993, pp. 121-142.

6 Leinfellner W., "The change of the concept of reduction in biology and in the social sciences", in: *Centripetal Forces in the Sciences*, Radnitzky, G. (ed.), New York, 1989, pp. 55-77.

7 Leinfellner, W., "The new theory of evolution - a theory of democratic societies" (in this volume).

8 Leinfellner, W., "The new theory of evolution - a theory of democratic societies" (in this volume).

REFERENCES

Bohm, D. [1984], *Wholeness and the Implicate Order*, London: Ark Paperbacks.

Carrier, M., Mittelstraß, J. [1991], *Das Leib-Seele-Problem und die Philosophie der Psychologie X*, Berlin: de Gruyter.

Erpenbeck, J. [1993], *Wollen und Werden*, Universitätsverlag Konstanz.

Geyer, F. [1991], *The Cybernetics of Complex Systems*, Salinas, CA: Intersystems Publications.

Götschl, J. [1991], "Hypercritical physical realism and the categorial changes in the subject-object relations", in: *La Nuova Critica*, pp. 5-19.

Götschl, J., Leinfellner, W. [1994], "Erwin Schrödinger's world view: the role of physics and biology in his philosophical system", in: *What is Controlling Life? 50 years after Erwin*

Schrödinger's What is Life?, E. Gnaiger, F.N. Gellerich, M. Wyss (eds.), *Modern Trends in BioThermoKinetics*, Vol. 3, Innsbruck: University Press, pp. 23-31.

Götschl, J. [1989], "Philosophical and scientific conceptions of nature and the place of responsibility", in: *La Nuova Critica*, pp. 5-22. For an abridged version of this paper with stronger emphasis on the coupling of material and non-material structural levels, see: Götschl, J. [1990], "Philosophical and scientific conceptions of nature and the place of responsibility", in: *International Journal of Science Education*, pp. 288-296.

Götschl, J. [1992], *Erwin Schrödinger's world view: the dynamics of knowledge and reality*, Dordrecht: Kluwer Academic Publishers.

Götschl, J. [1990], "Zur philosophischen Bedeutung des Paradigmas der Selbstorganisation für den Zusammenhang von Naturverständnis und Selbstverständnis", in: *Selbstorganisation - Aspekte einer wissenschaftlichen Revolution*, Krohn, W., Küppers, G. (eds.), Braunschweig: Vieweg, pp. 181-199.

Götschl, J. [1986], "Zum Subjekt-Objekt-Problem von transzendentaler und evolutionärer Erkenntnistheorie", in: *Transzendentale oder evolutionäre Erkenntnistheorie*, Lütterfelds (ed.), Darmstadt: Wissenschaftliche Buchgesellschaft, pp. 285-306.

Hörz, H. [1994], *Selbstorganisation sozialer Systeme*, Lit Verlag, Münster/Hamburg.

Jantsch, E. [1979], *Die Selbstorganisation des Universums*, München: Deutscher Taschenbuch Verlag.

Luhmann, N. [1985], *Soziale Systeme: Grundriß einer allgemeinen Theorie*, Frankfurt: Suhrkamp.

Nicolis, G., Prigogine, I. [1977], *Self-Organization in nonequilibrium systems: from dissipative structures to Order through fluctuations*, New York: Wiley.

Nicolis, G., Prigogine, I. [1989], *Exploring complexity systems*, New York: Freeman.

Penrose, R. [1989], *The emperor's new mind*, Oxford: University Press.

Prigogine, I., Stengers, I. [1984], *Order out of chaos*, New York: Bantam.

Riedl, R., Huber, L., Ackermann [1991], "Rational versus ratiomorphic strategies in human cognition", in: *Evolution and cognition 1*, pp. 71-88.

Schrödinger, E. [1958], *Mind and matter*, Cambridge: University Press.

Searle, J.R. [1992], *The rediscovery of mind*, Massachusetts Institute of Technology, Massachusetts.

Whitehead, A.N. [1967], *Science and the modern world*, New York: The Free Press.

Whitehead, A.N. [1978], *Process and reality:an essay in cosmology*, New York: The Free Press.

Wuketits, M.F. [1984], *Concepts and approaches in evolutionary epistemology*, Theory and Decision Library, Boston: Kluwer Academic Publishers.

Zeleny, M. [1985], Spontaneous social orders, in: *The science and praxis of complexity*, The United Nations University, pp. 312-327.

TOWARDS A SCIENCE OF MAN

OSWALD SCHWEMMER

THE ENIGMA OF MAN

What or who is Man? There is no lack of answers to this question. A special discipline became established in this century dedicated to answering it: philosophical anthropology. But despite this, and with each new answer that has been given, the question itself has become more and more mysterious. One widely entertained answer consists in declaring the enigmatic nature of this question about man's own nature to itself be the decisive answer. Man, we can then say, is that creature which remains a mystery to itself and which must remain so, because it cannot penetrate behind its own questioning or discoveries, since the spheres of meaning that it opens up with its questions extend beyond its own possibilities for understanding. The mind which it inhabits is more comprehensive than the particular thoughts which it can have.

We find that answer in classic German philosophy, the philosophy of Idealism, and it was given an explicit formulation particularly in Max Scheler's famous book about man's place in the cosmos.[1] At first glance this answer could be taken as mere escapism, as an attempt to avoid detailed conceptual effort by retreating to mere gesture, calculated to discredit all such efforts as mere half-truths, assembled in the business of empirical study. The particular answers of, say, the biologist, the historian, or the all too ready answers of philosophers, when seen from the viewpoint of this gesture, seem to take cognizance of only some detail, an aspect, a few splinters of a mosaic which is never complete. The whole is only found in the mystery, the riddle, which permits no solution. That is man.

FUNCTIONAL CIRCLES: BIOLOGY'S ANSWER

In reality, the situation was and is more complicated. Max Scheler, whom I see as one of the founders of philosophical anthropology in the modern sense, wanted to take account of the results of the sciences of man and base philosophical reflection upon them. Philosophy is supposed to pose the old question "what is man?" not just in cooperation with the modern sciences but also to take them as its point of departure: what is man, given that he is in any case an organism, a creature with kinds of perception and behavior? The answer to this newly formulated question turned out to be different, however, than expected. The biologist's answer to the question let the philosopher, so to speak, fall into the arms of the biologist. Biology appeared

J. Götschl (ed.), Revolutionary Changes in Understanding Man and Society, 129–146.

for the first time in the history of the sciences to be able to comprehend man - in the functional circles of perception and behavior, of recognition and action, such as Jakob von Uexküll delineated them in so fascinating a manner.[2]

This attempt to comprehend man in biology is based upon a differentiated understanding of the "surrounding and inner worlds of animals." According to this view the complexity of animal perception and behavior - and especially the puzzle of the instincts - could be solved by differentiating between particular functional circles in which a receptor world was opposed in each case to a different effector world. A frequent example was offered by the receptor world of the tick, a world constructed solely upon the basis of sensations of light and warmth and in which otherwise only the odor of buteric acid occurs, so that the mammals can be identified that the tick falls upon. A higher development in the animal kingdom means then a higher complexity in the worlds of sight, hearing, and touch, in the worlds of smell and taste. These combine to form a multi-form environment of numerous partial worlds, which are not always combined, in which the animals behave differently as they move from one partial world to another.

THE UNITY OF THE WORLD OF OBJECTS

With man, this variety is conjoined together to form a unity in which the many worlds form one continuous world. In this world what we see can also be heard and touched by us, or, to put it differently, in it we grasp the identity of objects which we sense with a variety of senses. That this is not a matter of course is shown by the many examples of separated sensory worlds in animals. From Hans Volkelt we have the example of the spider which feels the vibrations in its net when a fly is caught in it and which approaches the fly as its prey in order to paralyze it, to take it into its lair and devour it. Yet the same spider flees when the fly is put before its eyes.[3] Perhaps even more impressive are Köhler's observations of his chimpanzees, to which a first, primitive use of tools is attributed. In fact, their behavior is only guided by optical qualities, hence they use the bamboo stick and the rope in the same way in order to reach the banana hanging from the ceiling, even though the rope is only hanging down loosely. Or they attempt to stack boxes on top of one another according only to optical considerations - setting them just upon their corners - without paying any attention to static considerations, just as in the attempts to "stick" boxes onto the wall, in order to climb up on top of them.[4] Here the worlds of touch and of vision remain separate. Here, what is felt by tactile sensations is not recognized to be the same as that which is seen. And so biology appears capable of telling us what is characteristic of man and what determines his nature: the complex variety and unity of his surrounding worlds in which he alone can move and act as in a single world. But what constrains him, man, are the functional circles which also guide his behavior. Granted, these are more diverse and complex than those in the rest of the animal kingdom, and so the impression of independence can arise, which in the tradition was taken again and again for freedom. But in fact man remains bound within the spheres of reception and action, of perception and behavior, subject to the same laws as the animals.

MIND AND MAN'S OPENNESS TOWARDS THE WORLD

It was this biological approach to man that Max Scheler opposed. No, man is not bound within the functional circles of perception and action, no matter how complex and differentiated these are supposed to be. Man, Scheler contends, is the creature who is open to the world, who is able to break through these functional circles and who is able to grasp the unity of his experiences, of the objects of his experience and their all encompassing world due to the fact that he can set up a self-produced inner world in between the perception of the outer world and his action, an independent and wholly new world of the mind.

This mind is not to be confused with goal-directed activity, with technical activity, which is something that animals, such as Köhler's chimpanzees, are capable of engaging in to a limited extent. Scheler expresses this in a drastic remark: "Between a clever chimpanzee and Edison, taking him only as a technician, is only a - of course very great - gradual difference."[5] The mind is the capacity to determine the features of objects; it can conceive something as identical in all of the different spaces of its various surrounding worlds, and it can imagine worlds on its own, which it can oppose to the existing world and in this way break out of the strictures imposed by the closed functional circles. But this liberation has its price. Man, who can create a world in his mind and who can hold onto objects in thought, has fallen out of the secure circles of perception and action and has lost the security they provide. Now he only has himself to rely upon; he must stand on his own if he is to have a place at all.

As a result, the attempt to found a comprehensive science of man which also carries on the philosophical tradition has, for the moment, failed. One can read Max Scheler's final interpretation as an assessment of the fundamental irrelevance of the sciences for a full understanding of man, so that the sciences are consulted only in order to show more clearly their inappropriateness for the study of man.

SCIENTIFIC LIMITATION AND PHILOSOPHICAL CLAIMS

Today, in fact, this conviction still or again exerts influence, and not just upon the opinions of the philosophers, but upon scientists themselves. Scientists who study particular human proficiencies or attributes emphasize repeatedly the limited scope of the objects and results of their research. Not scientists but philosophers are usually, if not only, the ones who want to celebrate the triumphal solution to the enigma of man, and they are the ones who, in addition, have done so by calling into being new disciplines such as "neuro-philosophy".[6] Naturally, we can also celebrate too and proclaim the final explanation of the human mind by means of algorithms and computer models. But that way we would only be repeating a solution that has been offered over and over again for a long time as a method to answer questions concerning man. First, one narrows down the question more and more until finally a little answer suffices. This, of course, is not an answer to the original question but a way to dispel it. But this is also good enough for some, if they are theoretically modest enough.

RESCUING THE MIND

Before we dispense in this way with all common sense, it would pay to once again reconsider Scheler's answer and to ask if we know more today and can therefore answer differently than he did then.

Max Scheler wanted to rescue the mind. He thought that the notion of closed functional circles could never explain the mind's flexibility, especially its capacity to identify objects in the flood of our perceptions. So he insisted upon man's openness to the world, which is expressed in his power to establish fixed points of reference. The world is open on all sides without limits because man can give it a structure - the structure of objects, i. e., of things and qualities - which gives us a way to grasp things and to move ahead into ever more remote reaches of experience, ultimately reaching regions accessible only through scientific instruments and which, insofar as they are within the limits of perception at all, can only be perceived through models.

MAN AS AN ORGANISM

The more closely we consider things, we must grant Max Scheler one point, but we must diverge from him in another. What Scheler and whole generations of philosophers have overlooked for centuries in their desire to rescue the mind is the independent existence and claims of the body. Man is first of all an organism, a highly complex body in a world with other bodies. And this body is not just a shell, serving only to support and protect the mind, a shell which one could possibly put aside or which at least needs to be "deadened" by ascetic discipline in order for the true - namely immortal and spiritual - life to come forth. The body is the locus of our existence, which otherwise would be dislocated, nowhere. It is the locus of our relations to the world; in it the lines of influence coming from and going into the surrounding world are woven together again and again in a living pulse beat of coexistence and cooperation with all the bodies in the world.

THE BODY AS THE LOCUS OF OUR RELATION TO THE WORLD

Our body is our connection to the world, in a double dimension and in two directions. The first dimension is that of perception and action, of lived experience and expression. In this dimension our body grasps what is happening in the world and at the same time impresses its answer on the world. To have no body - this is a state which only philosophers and, before them, theologians could conceive of - and to have no body would mean to be without experience and without expression, without perception, with no behavior, without feelings and without a world, and so with nothing at all. And so we come to the second dimension of our bodily relation to the world, our dependence upon the world and our benefit to it. I do not need to prove our dependence upon it. This confronts us dramatically everywhere that man does not heed it, so that all those who live in a poisoned or dying environment themselves are poisoned and become sick. That we benefit the world too is not so obvious. But this too is true. Man is also a part of the world, who contributes to it with his bodily existence. Like the other animals he too is embedded in the many cycles and lines of influence in nature, which he also takes up and to which he responds. But with man a new form of receptivity and a new way of directing influences comes into the world, which makes new developments possible, but which also, as we know all too well, also hinders many other developments.

SOLIDARITY WITH THE WORLD

If we look for a designation which can cover man's situation, of his being bound up in the world, and which does justice to both dimensions of his relationship to the world, then this is the notion of solidarity. Solidarity with the world as a whole, as a thing among things, a living creature among living creatures, a body among bodies, this is the existential situation of man. At the same time it is a situation that man himself, in his history, has forgotten and repressed.

SELF-IMAGES OF MAN: DIFFERENCE INSTEAD OF SOLIDARITY

Man and the earth, that is from the beginning a relationship of dominance, of man's power over the earth, and of possession. In this self-conception man stands above and beyond everything else in the world. He lives of course in the world - but only with his body. With his mind, which only he possesses, he does not belong to the world, he transcends it and is close to God.

This deep line of demarcation running through our universe - on the one hand man living as the creature of spirit and on the other, the rest of the world including the bodily side of man - hampers not only our view of this world, in which man lives as an insurmountable environment, it also hinders man's own view of himself, who remains a bodily creature. Instead of his solidarity with the universe, a difference is put forward, a fundamental difference which is actually dominance, because it is guaranteed ultimately by God. This makes man a creature without a world and also without feeling, hence it makes him a kind of alien element who then haunts an extraneous world.

THE SCIENCE OF MAN: AN ECOLOGICAL SCIENCE OF SOLIDARITY AS A HUMAN FORM OF EXISTENCE

We have only begun to understand these interconnections. They demand a reversal in familiar and deeply rooted conceptions by which we have come to see ourselves and the world as disparate realities. Philosophy is called upon here to make this reversal intelligible, to rework the conceptual fields by which we understand our existence. In particular, it needs to show how the separation and opposition of man and world, which the philosophers usually base upon the opposition between mind and nature, is a separation of worlds - namely between those of the mind and the body - introduced by man himself, but making it impossible for him to understand himself.

THE MIND AND ITS STRUCTURE

With this we have come to the second point, the point where philosophical anthropology can assert its claims, its insistence upon the special character of the mind and, of course, reference to mind in the attempt to understand the essential nature of man. Here, it seems, a philosopher like Max Scheler has seen something that we cannot deny. For there is in fact a difference between the mind and the closed functional circles of the other animals, a difference which cannot be explained by a greater complexity and further differentiation of these functional circles. This difference concerns the elementary, and for us so matter of course, capacity to conceive of objects. This is something which seems to be simple. We can understand this in a preliminary way simply as the capacity to grasp the identity of objects - in connection with and in contrast to the impressions we have. Put differently, it is the capacity to understand all of our impressions immediately as aspects of objects, which as such possess their own identity.

THE IDENTITY OF OBJECTS AND THE WORLD OF OBJECTS

What for us is a matter of course is totally inaccessible to animals. Naturally, our dog remembers us, even if we have been gone for a long time. And in this sense he identifies us. But this identity is not the identity of objects. This recognition remains bound to the perceptual and behavioral situations that the dog encounters. The garden that he daily traverses will never reveal to him his paths in an overview. These paths appear to him only in the course of his movements. They arise so to speak and never permit him to obtain an overview of them or even to grasp them in a schema. They remain aspects of his experience from which he cannot extricate himself or rise above. The identity of an object however transcends the limits of situations following one upon another. The object - no matter how practical and trivial it may be, like a chair, a table, or the lectern I am standing at - retains its identity in the change of situations. More important and essential: objects are able by means of this, their identity, to spread a kind of net in which the changing situations can be recognized and are related to one another in the midst of change. Through the identity of objects we attain a new standpoint for our world experience as a whole. Beyond the changing situations we erect an interconnected and ordered world of objects in which we have now taken up residence and from which we observe the stream and course of events - to which also belong our own immediate experiences - and which we then can permit to enter into our perceptions or not.

EXTERNAL IDENTITY

This new basis lies outside of our immediate experience and perception. This is the decisive point. It cannot arise therefore solely from a correlative restriction of the different areas of perception. Such a restriction, no matter how complex it may be, does not lead beyond itself. It only offers internal orientations, i. e. internal to our experience and action. External identities, by contrast, require a medium existing and acquiring its structure outside our experience, an external medium. Max Scheler and subsequent philosophical anthropology recognized the external existence and structure of identical objects, what they overlooked is the fact that a medium, a material medium, is required so that such an external existence and structure can take shape. These authors did not grasp the "exteriorization" which concerns us here as a real process in our world, but as a product of thought in the realm of pure thinking, something beyond our world, not an existing thing, but a kind of real ideality - as one says in the language of philosophy. The mind for Scheler is a "world eccentric" being,[7] which, accordingly, makes man the "creature who can say no", the "ascetic of life", the "eternal protestant against all that is merely real".[8] If we recognize, however, to put it in a simple common sense way - from nothing can come nothing - then we have to look for the external medium in which the identity of objects can find their real basis.

THE MEDIUM OF SYMBOLISMS

This medium is provided by symbolisms, especially our language, but also by the worlds of images, sounds, forms of movement and of taste which people have developed, followed, added to, and changed in their cultures, or, to be more precise, which form the underlying basis and connecting medium of their cultures. With this view of cultural symbolisms a fully new perspective opens up upon man and the old question of the essence of man.

IMAGINATION: THE FIRST SOURCE OF THE MIND AND THE BEGINNING OF EXTERIORIZATION

Permit me to develop this perspective a little further. First, we need to take a step backwards, before the development of these symbolisms, to a second source of the mind: to the imagination. It is the imagination which first breaks through the functional circles of perception and action, and with it the rise of the human begins. At some point, an independent realm of impressions, feelings, and images must have taken shape in which the feelings arising with perceptions and movements no longer formed only the bridge between perception and action but rather now positioned these feelings as consciousness grew, so as to form a stationary reality while aimlessly circling about. This new reality owes its existence to the transformation of felt urges into a mental image, the continuation of our lived experience in images, which remain with us and can be recalled over and over again.

THE FLOOD OF IMAGES AS A VISITATION

And so we have the imagination, fantasy at the threshold of consciousness with which a new realm of thoughts and images come into play. At the beginning of consciousness these streams of images are not just a blessing. They are also disquieting. They are not just a liberation from life enclosed in functional circles. They are a kind of visitation upon man, against which it is necessary to struggle in order not to succumb. We can envision only with great difficulty this new situation in which fantasy is born. We come closest perhaps when we recall the ideas that befall us when we are alone in the sleepless darkness of a dreadful night, in which our standards of judgment have deserted us and our thoughts take on a life of their own and perhaps take the form of nightmares. Then it is sometimes difficult for us, the enlightened administrators of our thoughts and memories to draw a line between fantasy and reality. How must it have been then for those men whose consciousness was only dawning!

THE NECESSITY OF EXTERNAL SUPPORTS

Supports must be found to provide stability in the streaming flood of images, which at least permit us to give them partial guidance. These supports can only come from without. They do not reside in the images themselves, which only merge and displace one another, but which can neither provide limits for one another nor establish firm identities which can be referred to again and again. The external aspect of images is our body and its movements, our actions. These acts may at the outset not be much more than certain movements such as stamping with the feet or dancing, or shouting, or - if we can use this expression - singing. These are movements that are really performed "outside", that are conjoint and even identical movements, made on occasions that recur, perhaps only because they waken an urge to repetition, but also because they accompany a certain situation and finally come to characterize it: preparation for the hunt or the battle, a part of the celebration or invocation in time of need. Images attach themselves to movements, and especially to gestures and sounds. They are called forth by these movements and sounds or repressed, set aside, or remembered. In this way, with growing awareness, consciousness is no longer alone with the flood of images, and a kind of fellowship arises. Consciousness is being part of a group that dances and claps and stamps and shouts. The externality of these bodily movements, which is bound up here with the beating of a drum, is what tames the flood of images in the overwhelming inner worlds.

THE RITUAL ORDER OF FANTASY

Naturally the images remain and arise again. But they acquire new companions, companions from outside. The bodily expression of the imagination conjoins the flowing worlds of images with bodily movements which can then be performed again and again in the same rhythm or together in combined actions - in the same way as here and now. A ritual order of imaginative fantasy arises providing a first fixed point, a locus, that offers the possibility of something like a particular personal standpoint.

THE TENSION BETWEEN INSIDE AND OUTSIDE

Even here in the first dawning of humanity a tension appears which henceforth propels human development and yet at the same time makes this development more difficult to understand: the tension between inside and outside. This becomes all the more apparent if we consider the next important step, the acquisition of pure expression no longer bound to the whole body, but as something which arises in a medium of its own, a medium that belongs to it alone, and only for it, because it is created along with it. This pure expression is symbolism, especially the symbolism of linguistic expression.

THE UNIQUE STRUCTURE OF LANGUAGE

In order for a language to develop, more is needed than the particular sounds emitted by the speaker. Language is a unique system, a whole world of sounds, mixed with different bodily movements and states, a world which must already itself contain an order before it can be used for expression between speakers. I do not want here to tell any stories about the origin of speech. My only intention is to make language more understandable as a particular element of human life. This seems necessary not least of all because in the meantime we are confronted with an uninhibited and boundless use of language.

The many animal languages now talked about, like the now commonplace talk about computer languages have given the meaning of "language" a new sense and to an extent diverted our view from human language. This talk of animal and computer languages makes language a matter of calls and noises or algorithms and rule systems. Cries of pain and joy, growling and barking, advance to the prototype of language. And the computer's beeping corrections bring us possibly to the ideal of a thoroughly correct language.

What we forget here or at least have to forget is the fact that both - the animal's cries and the activation of computer switching - are signs of an elemental speechlessness. For what is a cry for us, a crying out, or - as the grammarian calls their domesticated form when it becomes part of our language - an interjection? They are the expression of an agitation, when we lack words. To be more exact, they are an immediate expression of our body rather than the result of thought. Ernst Cassirer put this pointedly, when he said: "interjections are employed only when one either cannot or will not speak."[9]. And so it is with the seeming ideal of computer language. In it we have formal transformations, which are error free in the particular given system, but at the cost of not being able to say anything. Logic remains tautological, and even where it is possible to compose formulations by means of inventive parallel switching circuits, they are based upon processes that follow from - and tell us what is built into - the programs. In order to understand human language and thus one of the essential aspects in man, we need to forget both the natural and artificial analogies and to direct our attention to what we really want to understand, namely to human language itself.

LANGUAGE A SYSTEM OF REFERENCES

Human language is an internally ordered system with internal references and not just a repertoire of cries, calls, or gestures. A language arises only when articulated sounds combine with one another to form a world with its own laws. When utterances only accompany and add to an immediate bodily expression they remain expressive splinters, arising in the situation, but not coming together to form an image. They possess their own particular relative immediacy in the situation and require no history in order to be understood. This changes with the rise of language. Language is from the outset a medium, a realm of its own, of ordered articulated sound, which transmits our expressions.

THE DIFFERENCE IN THE INNER WORLD OF MAN

This existence as a medium, the structure of language, introduces a difference into the inner reaches of humankind itself. For when we want to express ourselves, we must do so through work on a Gestalt which must be shaped in a medium with its own laws. Language is a living material, which has a life of its own, into whose laws and rhythms we must enter if we are going to say anything at all. How much effort it often costs us to find the right expression! This struggle for the right expression shows us with graphic immediacy the loss of immediacy.

THE MESHING OF THE INNER AND OUTER WORLDS IN CULTURE

But we must also know that this loss is coupled with a gain, which we can subsume under the general heading of "culture." Culture begins at the point where man first goes beyond his organism and constructs a world between individuals, a world of material signs and devices, which from now on leads and influences both the external world of human actions and the inner world of thought and feeling, of experience and striving. To put it schematically, we can say that culture consists in the formation of the inner world out of a self-produced outer world - as the inside of the outer world - and the formation of an outer world as a presentation of the inner world, as the exterior of the inner world.

Let's follow this meshing of inner world and outer world in culture more closely. A mere cry is the crying out of our body. It shoots through our consciousness so to speak, or, more correctly, it shoots past it. If we were to attempt to articulate it, we would block it. For the thing that is quite literally brought out in such an immediate utterance is a bodily state, a bodily feeling, which expresses itself in our body, in its crying out or other reaction, so our language here is like a discharge of energy.

If we were to monitor this bodily state or bodily feeling to find an articulation, it would disappear from view. Articulation is something else. Articulation consists in the symbolical fixation of the different regions of our thoroughly bodily inner worlds, so that they enter into mutual combinations. Hence, they are not only brought into their own symbolically articulated form, they also arise by virtue of relationships, which are possible only by means of symbolism. There is no such thing as an articulated inner life, about which we either remain in profound silence or which finds expression in speech. Often, whoever refrains from speaking has nothing to say. And whoever speaks or expresses himself in any way thereby attains with this actual expression the form and content of what it is that he wants to express.

The reason why there is no "pre-existence" in man's inner life of what is to be expressed is that our inner life - that which tradition refers to as the human soul - can only arise through the stabilizing interconnection of his partial worlds and this combination itself is but the work of the symbolization of our experiences.

This is what makes our world into the human world of the mind or, as we may also say it, of culture: that between the partial worlds of our senses, the worlds of hearing and seeing, fixed and reusable connections are created, bridges through which a whole world can arise from these partial worlds and in which all the different sensations from the different partial worlds come together in the sensory experience of objects. And because these "bridges," namely the stabilization in symbolisms of our different experiences of things outside our organism, are established in their own intermediate domain of symbolisms, they do not fade away into and with the rhythm of our organic existence. They remain as a self-contained realm - an outer, intermediary realm in and through which human life finds expression.

THE INNER WORLD IN THE SYMBOLIC INTERMEDIARY WORLD

In other words, man develops his inner life first in the nonorganic intermediary world of symbols. Symbols are like crystallization points around which meaning and form, expression and understanding grow and which provide starting points for future attempts at expression and understanding. Culture arises in the intermediary world of symbolism, in which our mind can take shape and define itself. The mind is not some inner life, which takes shape by and for itself and only later in an incidental way seeks after expression. The mind is a multi-dimensional formative process in the intermediate realm of symbolism, a process which takes shape in the living material of these symbolisms. The mind, so we may also say, possesses an intermediary existence between individuals, i.e., a cultural existence.

THE CONSTANT INNER REALITY

I do not want to suggest that there is no enduring inner reality at all, possessing its own being and remaining unaltered through all of its attempts at articulation. We have this enduring reality in our bodily existence with its inner feelings and - to an extent - in our imagination too as we experience it most vividly when we are alone, such as in our dreams or fantasies. Both our bodily feelings and our imagination have their sources in our body, part of which is the brain. Both are of course already impregnated by our symbolic forms of expression and understanding, although our bodily feelings much less so in their vague massiveness than are the more pregnant, but also unbounded, ideas of our imagination.

THE SCIENCE OF MAN: AN ECOLOGICAL SCIENCE OF THE SYMBOLIC FORM OF HUMAN EXISTENCE

The inner life and outer world, consciousness and being, thought and speech - all of these distinctions which belong to the basic inventory of our traditional conceptions of the self and the world are on this view in need of revision. Most of all, they need to be set aside as lines of division that isolate man's intellectual existence. Here too we need to formulate a postulate that can guide a future science of man: mankind must be understood in terms of the symbolic form of human existence. His intellectual achievements must be recognized as individual, or to be more precise, individualizing processes in which an inter-individual "material" is given form, namely the form of cultural symbolisms. But this "material" is not something given or existing outside of his activity. It is the "material" out of which he himself is made. Understanding and thinking, conceptualization and awareness, articulation and silence - these are concentrations in symbolic fields and worlds, which provide an inexhaustible environment for our consciousness or mind. Our mind exists only in this symbolic environment, as part of it, as a magnifying glass or mirror, taking it in and leaving its imprint on it at the same time. Here too we must therefore again speak of an ecology, but this time of an "ecology of the mind,"[10] which investigates the correlative relations between the different particular intellectual achievements and their symbolic environments.

While we can only speak of such an ecology of the mind as a program, rather than as a reality among researchers, we can nevertheless at least sketch some of the consequences which the realization of such a program would bring, its theoretical and practical consequences.

THEORETICAL CONSEQUENCES: THE METAMORPHOSES OF THE MIND IN THE TRANSFORMATIONS OF SYMBOLISMS

Let us begin with several theoretical consequences. The mind which no longer lives in an unhistorical inner world of pure thought and feeling, insofar as we attribute feelings to such a mind at all, now becomes a historical process gaining its existence in the construction and formation of cultural symbolisms. Like every other kind of historical process it can be empirically investigated and - like every kind of historical process - it is subject to change and so was not always what it now is, nor will it remain what it has become. We can make this situation clear when we follow the epochal transformations in forms of thought on the basis of the development of new symbolisms. Here it may suffice simply to point to some of the "metamorphoses of the mind" which have resulted from new media in the use of language, namely in the shift from merely oral forms of language to written forms (themselves quite varied), from handwritten copying to printing, from writing or drawing on paper to computer supported word processing and computer aided design.

In an oral culture, where people must rely upon memory when they want to report or tell something, other kinds of aids to memory must be used than in cultures where it is possible to write down or read what has been heard or is to be told. Rhyming syllables, alliteration, rhythmic and melodic division, repetition, fixed sayings, as well as thematic rhythms and much more are typical aids to memory for the narrator. These are familiar to us through our fairy tale literature, from the Homeric epics, and from the Bible.

Once writing stands at our disposal, then texts of any desired length and complexity can be produced, texts which exceed the bounds of our memory. Here it makes a definite difference which kind of writing is used, a pictorial writing (as for example some Chinese writing) or a phonetically interpreted writing, a syllable or letter-based writing, merely consonant or also vocal alphabetic writing. With the different kinds of writing, different units of meaning are defined, and different grades of context dependence created, and, most important, however, different interpretations occur. In the phonetic and, hence, language-bound forms of writing these are primarily found in different segmentations in the spoken language.

Writing binds language to two different media, which interpret, modify, and differentiate one another, and it opens up completely new possibilities for symbolic representation through this correlative relationship. That which we frequently take to be laws of thought - such as logical truths or the principle of causality, understood here in the general sense that it permits inferring the actor from the act - come about as the result of a "semantic mirroring" of the organizational principles or requirements for writing a spoken language. The linear regulation of expression, the sharp segmentation of grammatical (written) elements, the greatest possible clear delimitation of the units of expression (i.e. in sentences), the grammatically most complete composition of texts, units to which one can then refer in further speaking, writing and thinking - these are the kinds of principles and requirements that are mirrored in our reflections as laws of thought.

The introduction of book printing combined a new medium with language. This made possible the exact reproduction, e.g., of (mechanical) drawings, as well as the two-dimensional arrangement of texts and symbols and so a new combination of symbols, schemes, drawings and text. This new combination opens the text in particular to the visual imagination - be it that of the technician or of the artist. The text is no longer understandable by itself, no longer is it a closed system, providing a complete verbalization of the represented world. With this the forms of thought change again. The scholastic disputation moves almost exclusively in the world of texts, pitting argument against argument, piling them up and assessing them in this textual world, but now this process has been interrupted by the curiosity of observation, by the desire to grasp and observe, which here breaks free of the discipline of the logic of language. The result is the spirit of the modern mathematical and experimental sciences.

Today's new media are the fruits of this new start and in the modern age these previously dominant modes of thought appear to have come to an end. The final projection onto paper, and hence the linking of language to two-dimensional representation - even when it is avoided through the (quite complicated) construction of models and technical practice, is reduced now to an intermediate result of the computer. And while at present everything still is printed out onto paper, some of the printed symbols can only be resolved on the screen.

Even though writing hardly becomes superfluous because of this, the computer poses a challenge to its domination of humanistic scholarship, where, until today, its rule has been practically unopposed. We today are witnessing typical transition syndromes which on the negative side appear as an atrophy of the ability to express oneself in written language and in an uncontrolled media consumption, but on the positive side appear as a new openness for changing perspectives and so also for the viewpoints of others, an undogmatic, pragmatic attitude even in questions concerning principles, and so - at least as a chance - to gain a better overview of complex decision situations. Another metamorphosis of the human mind has begun and with it another stage in the life of our culture, a stage in which we have begun to organize our intellectual life, or as we can now also say: our symbolic existence.

A glance at these metamorphoses of the mind lets us recognize their interwovenness, in which the symbolic media, cultural symbolisms, human intellect, and each particular culture form a whole composed of many voices to which we ordinarily refer with the global designation "mind" or the "culture of an epoch."

PRACTICAL RESULTS: THE MANIPULABILITY OF THE MIND

Now we come to the practical results of this view of man. If the symbolic constitution of the human mind gives it a historical, material shape, then - like all other material things - it is subject to degeneration and manipulation as well as receptive to the will to form and construction. The extraorganic intermediary existence of the mind makes the mind, at least in part, a kind of object, an object that we can preserve or destroy. To put this in a formula: the symbolic-cultural aspect of the human mind is in our hands and is entrusted to us at the same time.

That this is not an exaggeration is shown by historical examples from the recent past and present day. Ernst Cassirer attempted in his book on the Myth of the State to show how in National Socialism the material of the mind was actually misshaped through an entire "technology of myth". I mention his investigations here only as an example of an attempt. (We cannot yet speak here of the results of such investigations, because there are to date no reliable lines of inquiry on the subject.) Cassirer sees the "technology of myth" as it was developed in this century as a devastatingly effective instrument for the collective manipulation of consciousness. To this technology belong: (1) the manipulation of language, limiting communication and restricting criticism by means of a semantic emotionalization, (2) the ritualization of action so as to level off the difference between the private and public sphere and so to rob people of their individual possibilities of resistance, (3) the elimination of general ideals and the replacement of conceptualization with concrete images, which are supposed to obstruct critical distance, and (4) the reinterpretation of history as fate, which relegates the task of interpreting history and especially of the future to "providence" and to its prophets.[11]

Even though this diagnosis surely underestimates the role of "real factors" and overemphasizes the role of the manipulation of symbolism for the collective deformation of consciousness, it nonetheless makes something clear which all too easily disappears from view in the traditional understanding of the distinction between the inner and outer worlds: that the historically developing reality of the mind exists in an intermediary world between individuals, where it is accessible and vulnerable, and that it is not the inner primordial creation of lone individuals (in German culture these are the "Dichter und Denker," the poets and thinkers), rather the living principle and signature of a cultural situation, to which each and every symbolic achievement in a particular time belongs and contributes.

MAN AND THE AIMS OF A SCIENCE OF MAN

We could pursue further consequences of this view, but I will return here to our initial question about man and to the sketch of a program for the science of man. What is man, and what can a science of man be like, and why do we require answers to these questions?

I will press the first answer into formulas that look like definitions. Man is an organism existing in solidarity with and in the natural world. In addition, man is an intelligent creature, possessing a symbolic existence between individuals and in the cultural worlds of society or a plurality of societies. His identity derives from the individuation of organic and symbolic forms, but this individuation is bound to his physical and symbolic surroundings, in and through which he gives shape to himself as a "living material." The freedom of individuation and the independent being of the structures in man's surroundings stand in a tension, both poles must be recognized and reckoned with, but they must never be divorced from one another or taken by themselves as the definition of man. This applies also to the tension between man's organic and symbolic existence, which intersects, so to speak, with the tension between individuation and the structures in man's surroundings.

For a science of man, it follows that it needs to investigate the interconnections and tensions between these poles and fields. I already characterized them as an ecological science in a twofold sense: as a natural ecology and an ecology of the mind. In each of these two conceptions, however, it is a science of the correlative relationships between the developing organic and intellectual identity on the one hand and the natural and symbolic surroundings of this process on the other hand. What is not included in this - and possibly cannot ever be included in it because it is inaccessible to scientific method in the sense of theoretical rationality - is individuation. Individuation, ultimately, is not subject to theoretical explanation insofar as it is an act of freedom which finally determines the ultimate shape of intellectual and organic identity. One of the tasks of a science of man which does not hide behind its method in a narrow-minded attitude about spheres of competence is to recognize the threats to the possibility of individuation and to identify and to preserve these possibilities and to provide them where there are none. It needs in other words to keep open our view of man's entire sphere of existence - even for what is found at most only in art, but in no case in such a science itself. We know therefore only too well why we need a science of man.

Prof. Dr. Oswald Schwemmer
Institut für Philosophie
Humboldt-Universität zu Berlin
Unter den Linden 6, 10099 Berlin, Germany

NOTES

1 *Die Stellung des Menschen im Kosmos* appeared in 1928 as Scheler's last book. It is cited here according to the new edition in the *Gesammelten Werke*, Band 9: Späte Schriften, Bern and Munich 1976.

2 Jakob von Uexküll, *Umwelt und Innenwelt der Tiere*, Berlin 1909, (second edition, 1921); and *Theoretische Biologie*, Berlin 1920, (second edition 1928; Neuausgabe: Frankfurt/Main 1973).

3 Hans Volkelt comments upon this observation this way: "In the consciousness of the spider the fly is not represented by the same thing-like structures as it is outside of them... The constancy of thing-like structures is found in only a very narrow sphere of situations." *Über die Vorstellungen der Tiere*, Ein Beitrag zur Entwicklungspsychologie, Leipzig 1912, p. 56.

4 Cf. on this the chapter on the "Leistungsgrenzen der Tiere", in: Arnold Gehlen, *Der Mensch*, Seine Natur und Stellung in der Welt, Wiesbaden 13th ed. 1986, pp. 149-157; American edition: "The Limits of Animal's Achievements", in: *Man*, His Nature and Place in the World, New York 1988, pp. 137-144.

5 *Gesammelte Werke*, vol. 9, p. 31 note 1. In fact, Scheler here underestimates the accomplishments necessary in order for tool use to come about at all. As the limits to the achievements of Köhler's chimpanzees show, it is necessary, in order for objects to be used as tools, that they first be recognized in several dimensions of perception - e.g. optically and tactically - as being the same and so as identical objects in different surroundings. The fact that something functions merely as an optical bridge between one's own standpoint and that of an object to be grasped does not make the thing in question into a tool. In contrast to Scheler's judgment, an act of objectification is first necessary, something which Scheler wants to reserve in mind as something which is above the technical.

6 This term has come into circulation since the publication of a book by Patricia Smith Churchland, *Neurophilosophy*, Toward a Unified Science of the Mind/Brain, Cambridge, Mass., London 1986.

7 Scheler, p. 69.

8 Scheler, p. 44.

9 Ernst Cassirer, *An Essay on Man*, An Introduction to a Philosophy of Human Culture, New Haven-London 1944, p. 117.

10 So the title of Gregory Bateson's book: *Steps to an Ecology of Mind*, 1972.

11 Ernst Cassirer, *The Myth of the State*, New Haven - London, 1946, Chapter XVIII: The Technique of the Modern Political Myths. See also John Michael Krois, *Cassirer*, Symbolic Forms and History, New Haven - London 1987, pp. 191-198.

PART IV

RATIONALITY, COMPLEXITY AND UNCERTAINTY: NEW INTERRELATIONS

THE NEW THEORY OF EVOLUTION - A THEORY OF DEMOCRATIC SOCIETIES

WERNER LEINFELLNER

1. THE TWO WAYS OF HOW TO SOLVE SOCIETAL CONFLICTS AND THEIR PSYCHOLOGICAL ROOTS

The figure below is a graphic abstract of this paper. The triangle shows the transitions from the biological to the stochastic and the sociocultural theory of evolution. Outside the triangle, some disciplines are listed which are today incorporated into the stochastic theory of evolution or possess similar structural features. The solid arrows connect the historical sources with later developments; the broken arrows stand for structural, analogical, and similar comparisons between the different theories of evolution discussed in this article.

Figure 1. RISK, UTILITY THEORY & THEORY OF GAMES
(Serial Conflict solving under uncertainty and risk)

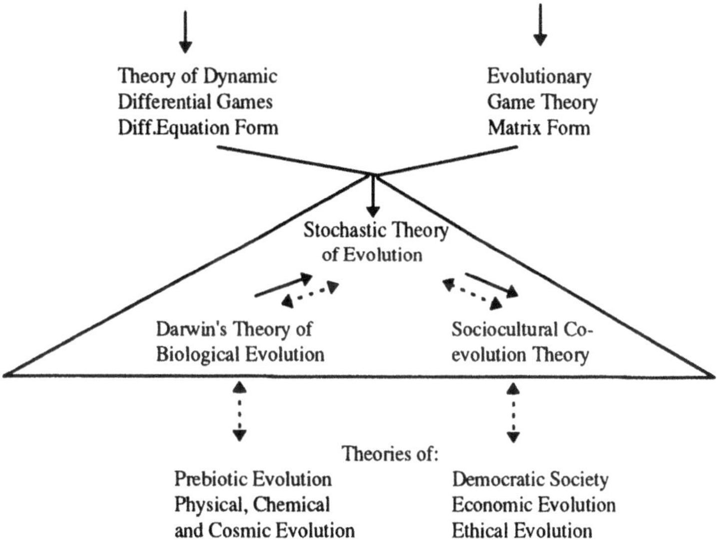

Since antiquity, philosophy and, particularly, 19th century social philosophy attempted to found the progress of societies on the evaluating, deciding, and acting individuals. In the twentieth century, psychologists and game theorists (H. Simon, A. Rapoport, W. Leinfellner) proposed a new name for

J. Götschl (ed.), Revolutionary Changes in Understanding Man and Society, 149–189.

this most important mental skill: intelligent problem or conflict solving, a new paradigm for this inner process, guided the pioneering works of von Neumann, O. Morgenstern, and Luce and Raiffa. The paradigm permitted not only the theoretical explanation and reconstruction of conflict solutions and crises, but also the computation of optimal solutions of serial social conflicts by new models and theories. Soon conflict solving turned out to be the most complex mental and practical human activity. One realized that this activity involves all the mental and practical skills which individuals possess. It involves preferences, evaluations, probable inferences, rules, storage of past experiences in individual and collective memories, and maximizing expected utility and social security - even under risk and uncertainty. Finally, the theoretical models showed for the first time that all these processes, especially serial conflict solutions, are controlled, and guided by, our innate risk behavior when we face random events. Today, the theory of conflict solving includes optimal solutions of evolutionary serial conflicts, their realizations, learning by repetition, and even creative decision making.

In the beginning, game theorists analyzed only single games in isolation. Later on, the "static" solutions of single conflicts became a special case of solving dynamic or serial conflicts. The word "game" became "synonymous" with "solving serial conflicts". Everybody knows that we learn best from iterating a game and keeping our good and bad experiences stored away in our memory in order to avoid repeating old mistakes. To adapt solutions by iteration and to improve them until they are optimal can only be achieved by learning from our past experiences; this comes closer and closer to evolutionary conflict solving. No wonder that game theory developed between 1944 and 1980 from a merely static theory to a dynamic one, and, finally, to the present evolutionary game theory (Mesterton 1992). Here we have a convincing example of a Kuhnian theoretical "exhaustion" of a prescientific paradigm by its reconstruction in models and theories.

Since probabilities and random events are involved, dynamic solutions are more complex than mere logical explanations. The decision maker may choose not only between the moves and strategies of a single conflict but also between possible serial conflicts. Moreover, besides the already-mentioned factors, the decision maker's solution depends on the other social participants, on random events, and on a set of socially accepted rules.

The dynamic differential game theory leads to solutions of serial games or conflicts. It shows not only how our intelligence may solve future conflicts in our imagination, in mente, but also how we can realize our solutions of societal conflicts. This unending conflict solving and its realization turned out to be the socially decisive theoretical as well as practical task. The gist of dynamic game theory provides us with a theoretical reconstruction of our individual and mental processes when we solve serial social conflicts - a rather unexpected outcome. Dynamic game theory reveals the psychological factors of conflict solving: firstly by showing its dependency on my internal, psychologically rooted preferences and evaluations which order and rank facts, events, things, acts, moves, and strategies - even those of all the other individuals. Secondly, it discloses the evolutionary function of our psychological risk factors (Allais 1979: 41-60). Thirdly, the new theoretical approach to conflict solutions showed for the first time how human beings represent external random events by internal probable estimations and risk, rather than by lawlike propositions. Fourthly, conflict solutions rely on our internal storage. We store past default rules and past experiences in our individual and collective memory and recall the batch of suitable rules to solve a specific conflict. Any evolutionary process - according to Hawking even the physical evolution - is a history-protecting or -preserving process. Fifthly, serial conflict solutions combine all these factors into an inductive algorithm which allows us to infer optimal solutions of evolutionary societal conflicts.

This highly important mental process which answers to the question of what we shall do or of how we shall optimally resolve a conflict has been reconstructed by two theoretical methods. One is the traditional evaluative matrix method, the other a new solution algorithm, known as "evolution equations".

2. THE EVOLUTIONARY CHARACTER OF SOCIETAL CONFLICT SOLVING

As soon as game theory became dynamic and began to deal with the solutions of temporally ordered series of games, it could also be applied to explain evolutionary conflicts, that is, serial conflicts whose solutions and batch rules are transmitted from generation to generation. These stored solutions gradually or abruptly stabilize serial conflicts, since they have an adaptive function. For example, they optimally adapt the course of evolution to the changing environment or to suddenly appearing selective random events.

If we want to gain a foreknowledge of the future, the game paradigm is very useful. To find solutions of complicated societal games or conflicts is the most complex intellectual task of mankind in general and of all sciences in particular, especially since we can't solve them by simply extending them like natural laws into the future. Future societal uncertain and risky conflicts show up in our minds not only as simple alternatives but also as compound alternatives of future simple alternatives. Later we will show that both can be represented theoretically as lotteries. Compound conflicts consist of conflicts of simple subconflicts where each subconflict is again a serial conflict. In our mind or imagination, a compound conflict always divides our future into two or more possible temporal iterations of subconflicts, just like a decision trees branches off into more and more branches. Subconflicts form evolutionary trajectories of solutions. In order to search for a solution between conflicting branches, i.e. subconflicts, we choose, in mente, among their possible future solutions. It is well known that a good chess player evaluates in his mind between two and three different trajectories (combinations) of his and his opponent's future moves and their possible outcomes before he finally realizes the - hopefully - optimal one. Each trajectory, then, consists of some of his own future moves and his partner's countermoves.

To sum it up: the first mental or internal step when solving a societal serial conflict consists in allotting to all open alternatives and subalternatives our individual preferences and evaluations. Usually, this involves uncertainty and risk. In a second step, we infer inductively the best possible solution by using accepted and stored societal rules - or rules of the game - for example, the maximization of expected utility. In the third phase, our risk attitude modifies the expected utility to a "risk utility"; for example, our psychologically rooted risk attitude modifies the maximization of expected utility. Thus, we obtain an optimal prima facie internal or mental solution. The last step is the realization of the optimal solution; at the same time, the realization is the solution's test and its confirmation by the empirical reality. Here we have a sort of paradox: it seems to be easier to create or realize our future than to predict it reliably.

Today's basic research in the field of utility theory and dynamic game or decision theory focusses, on the one hand, on this psychological and even genetically rooted interpretation of conflict solving. This will be demonstrated in the next chapter by the matrix method and by models of conflict solving.

On the other hand, the general differential equations of evolution (Eigen & Schuster 1979: 6-15) reconstruct in stochastic form the course (trajectories) of evolutionary conflict solutions under the impact of random events. They give us a perfect mathematical reconstruction of how we may calculate solutions, i.e., the future course of adaptive serial solutions of conflicts. They fail, however, to illuminate the psychological foundations of conflict solving or decision making under uncertainty and risk.

3. THE GOAL AND PURPOSE OF THIS ARTICLE

This article rests on one basic hypothesis: our intelligence as well as the intelligence of higher animals is an evolutionarily acquired, psychologically and genetically rooted mental and practical ability to solve serial conflicts; this ability functions as an optimal adaptation; it has been "designed" by evolution to cope with the advantages and disadvantages of unending conflict solutions, especially under the impact of uncertainty and selecting random events; it is guided by our risk attitude. This article will show how, and to which degree, human societal evolution depends on our intelligent and practical skills to solve the never-ending sequences of evolutionary conflicts and to intervene practically in the course of evolution.

Today, this hypothesis can be vindicated scientifically within the framework of the general or stochastic theory of evolution, a new supertheory which is just emerging from dynamic game and utility theory (Fig. 1). It developed in two steps: firstly, by melting together various theories, such as utility theory under uncertainty and risk, with evolutionary game theory in matrix and equational form. The method of evolutionary game theory is similar to quantum-theory method. The dynamic in both theories (motion and evolution) is being governed by a system of generalized differential equations (Schrödinger's and Eigen-Schuster's equations). Eigen Schuster´s equation depicts the "cumulative result" of all factors of evolution, the matrix method the individual's and the group's influence on evolution. Secondly, this new theoretical supertheory grew by incorporating the new model of utility and of risk and other models of game and decision theory (e.g., n-person game theory, competitive and cooperative game theory, collective choice theory, statistical decision theory) into evolutionary game theory. Thirdly, the general theory of evolution reformulated mathematically, and incorporated successfully, the new versions of traditional principles of biological evolution: Darwinian adaptation and selection and the principles of mutation and survival of the fittest. Moreover, it took in anthropological and societal principles, such as history protection (Hawking's term), memory storage, the normative societal principles of welfare democracies, as well as democratic prescriptive rules of democracies. In other words, it replaced the older, vague concepts and principles by mathematical or stochastic formulations.

Within the context of the new stochastic theory of evolution, we are in a far better situation to show to which degree our intelligence can master not only our biological, but also our economic, political, social, and cultural evolutionary conflicts. If the successful solutions of evolutionary conflicts depend on the role of our intelligence, then this amounts to an explanation of the internal, mental mechanisms of how we solve and realize serial evolutionary conflicts under uncertainty and risk. And this can be taken as a proof of the psychological foundation of the evolutionary dynamics of economics, of political, social, and cultural science, insofar as they deal with the solutions of serial evolutionary conflicts. This is, in a - rather large - nutshell, the gist of this article. M. Allais and B. Arthur go so far as to regard economics as an intrinsically subjective science, hence a science which cannot predict.

4. THE PSYCHOLOGICAL AND DYNAMIC ROOTS OF THE STOCHASTIC THEORY OF EVOLUTION

As already mentioned, the stochastic theory of evolution describes theoretically and practically evolutionary serial conflict solutions with the aid of two different stochastic methods (Schuster 1985: 255; Leinfellner 1984: 248-246).

One is the dynamic matrix method which lays bare the changes of the evolutionary status from moment to moment; it explains the changes psychologically. The other one, the differential equations of evolution, represent phenomenologically and empirically the impact of random events and of temporal external changes during any evolutionary process. For example, it computes the influence and changes of random events on the next following solutions of conflicts.

It is very well known that the matrix is a value matrix and was introduced for this purpose by von Neumann and Morgenstern and by Luce and Raiffa. The dynamic value matrices span up an n-dimensional value-space in which the evolution can be described mathematically. In serial conflict solutions, each game matrix represents a cross section of each individual's expected evaluation of his chances in the future course of evolution at a certain time point. In brief, each matrix expresses the psychological, individual point of view of the evolution at a certain point in time. Thus, the changing value-matrices characterize the temporal trajectory of the ongoing evolution. The other method, especially the Eigen-Schuster differential equation of evolutionary changes, expresses the dynamic of evolutionary processes, in our case the temporal trajectories of solving serial conflicts. It permits the calculation of the stable or unstable outcomes of evolutionary trajectories which are formed by the series of conflict solutions.

To the great astonishment of game theorists, these two theoretical techniques which use entirely different methods and seem to have nothing in common yield the same solutions. This article will show - convincingly, I hope - that the two methods represent the dualistic aspects of social and biological evolution theory: the value and risk aspect and the phenotypical dynamic of evolutionary processes. The matrix method describes the psychological roots of evolution and evolutionary conflict solutions with the help of utility, value, and risk theory. The equational method describes phenotypically the temporal course or the dynamic of the evolution. The "evolution equations" are more general than the matrix method, since they can be used to explain even the propagation of prebiotic evolution, i.e., the evolutionary trajectories of complex molecules such as DNA molecules and enzymes (proteins), according to Eigen and Schuster. These entities are the most primitive units of life which possess no conscious preferences, feelings, or evaluations.

Interestingly, the dynamic matrix method begins with the individual's internal intelligent preferences, evaluations, and risk attitudes at a certain time which determine partially the future course of the evolutionary trajectory in question. Formally, this situation resembles a lottery, which we will call an evolutionary lottery. Of course, intelligence, which uses this type of lottery, does more than evaluate. Briefly, besides the disposition of perceiving, of sensing, of forming preferences and utilities (values), intelligence - in contradistinction to rationality - uses its conscious imagination, its individual and intergenerational memory storing faculties to master the future. They enable the individuals to solve optimally future serial conflicts as described in evolutionary game theory. Whereas the evolution equations can be used to explain already the prebiotic, molecular evolution (Eigen, Schuster), the value matrices are the solid method when we wish to explain the evolution of higher animals and the societal evolution of humans (Leinfellner 1984: 253-262).

In static game theory, the matrix method was used from the beginning to represent coherent psychological preferences, evaluations, and the optimal strategies of the agent's alternatives. In dynamic game theory, the evaluations yield a temporally changing series of interdependent value matrices at consecutive time points t_1, t_2 ,..., t_n which form the evolutionary trajectories. Each is a discrete description of the changing preferences and values during the course of evolutionary conflict solving. Thus, the temporal series of value matrices M_1 at t_1, M_2 at t_2, ..., M_r at t_r expresses discretely the changing evaluations at the temporal moments of evolution or of serial conflict solutions. The agent's changing preferences and values are partially caused by competitive or cooperative members of the society, by random events, by changes in the environment, etc. Whereas the temporal series of value matrices represent discretely the evolutionary paths, the evolution equations represent the evolutionary paths (the trajectories) in the value space of evolution in a continuous way.

5. THE BASIC UNITY OF SOCIETAL EVOLUTION

Just what it is that evolves has always been a mystery. Is it the genes, the individuals, or the societal forms that are the units or carriers of evolution? Here any unit or system whose dynamic can be explained by either one or by both methods is a unit of evolution. It is, then, no wonder that Lumsden and Wilson regarded customs, behaviors, institutions, artifacts, mentifacts (mental concepts) as units of the socio-cultural evolution (Lumsden & Wilson 1981: 368). Each of them is used by certain individuals and can be expressed by the frequency of its usage x_i within the population. Such units are societal forms. Because their function is similar to the biological function of genes, Lumsden and Wilson called them "culturgens" (sic).

Obviously, from a material point of view, culturgens have nothing in common with biological genes; but they have the same evolutionary function. Their frequency may increase or decrease in a population or it may remain stable, just like the frequency of biological genes varies within the genetic pool. Culturgens can be "inherited" or transmitted from generation to generation by learning, just as genes are inherited.

A simple example may illustrate the complexity of societal evolution. We can compare the evolution of any societal form to the way people live and act in a fortress. The people in the fortress are stratified into different societal groups or forms. For the sake of simplicity, we assume that there are only three societal groups in the fortress: defenders, neutralists, and pacifists. On all sides, the fortress is surrounded by high walls which secure its stability and the security of its mini-society, its population. The three societal forms have different preferences and evaluations; but they all strive for the greatest possible security. They strive more for the stability and security of their societal form or group than for their individual survival. In the course of time, conflicts will occur. Each societal form will solve the ongoing conflicts in its own way. Which societal form, i.e., the number of its members, will increase or decrease depends on random events, on external and internal enemies, and on the cooperation with friends. Societal and environmental random events may increase or diminish the number of people who belong to any of the three societal forms; but they all are guided by a hidden, psychologically rooted attractor: maximizing the security and stability of their society. From an evolutionary point of view, that societal form will increase its membership which solves optimally the constantly occurring conflicts. The course of evolution is characterized by the ups and downs of the societal forms. That societal form which improves the security and stability of society will increase, stay evolutionarily stable, will flourish and prosper, provided no disturbances by big random events occur. There is no guarantee that big events will not disturb this stability; chaotic changes will lead to the decline of the leading societal form. Possibly, it may even go extinct. The question is: which of the three societal forms in the fortress will increase, multiply, stay stable, or, put simply, in terms of Darwinian evolution: which one will survive?

6. THE DYNAMIC OF THE VALUE-MATRIX FORM

The dynamic matrix form reveals the psychological aspect of evolution, i.e., the likelihood of the expected risk value of each individual with respect to its future, as seen from its momentary point of view. It provides us with an elegant proof of the psychological and even genetic foundation of serial evolutionary conflict solving. This, in turn, has far-reaching consequences for the psychological foundations of the evolutionary dynamic of economics, of political, social, and cultural science. These two methods plus the principles of the general theory of evolution furnish us with a new theoretical framework which allows us to explain societal evolution as a constant struggle for societal stability, a struggle to solve the seemingly unending societal and environmental conflicts under uncertainty and risk. It replaces social Darwinism: for example, the Darwinian concept of an egoistic, merciless fight for the survival of the fittest is being put under the constraint of a cooperation which again is governed by democratic societal welfare principles.

Recently, traditionally "static" microeconomic and macroeconomic models have suffered setbacks, since they were unable to prognosticate the last recession. This shows that static models very often lose their empirical anchorage in the acting, deciding individuals, in their changing preferences, evaluations, and risk behavior; moreover, they neglect the randomness in our society and environment. The dynamic matrix method and the "evolution equation" method, however, include shifting disequilibria and equilibria. These methods permit to found economic, political, and societal sciences psychologically on preferences, evaluations, and risks of the individuals and groups. They also explain the economic dynamic evolutionarily by changing equilibria, by attractors, sinks, saddle points, all of which influence the trajectories of evolution (Eigen & Schuster 1979: 38).

7. TWO WAYS OF CHARACTERIZING OPTIMAL SOLUTIONS

Since the advent of the general stochastic theory of evolution, game theory cannot be considered any longer merely as a part of classical linear mathematics. Within the classical framework of mathematics, static game theory appears to be based on strategies, on the theory of static equilibria, on lawlike hypotheses (such as the optimization of utility), or solely on principles of competition. Therefore, the recent extension of evolutionary game theory to the stochastic theory of evolution marks a revolutionary turn to the dynamic of nonlinear systems. Step by step, mathematical-stochastic reformulations replaced the traditional concepts of adaptation, selection, of persistence (stability) of behavioral and societal forms (traits, customs, artifacts, and mentifacts), and of how to store and transmit optimal solutions and batch rules for finding solutions. The reformulation of the concepts of traditional evolution theory in evolutionary game theory permitted even the incorporation of most democratic welfare principles and rules into societal evolutionary conflict solutions. Finally, the new formulation of the dynamic of evolution by the equations of evolution yielded a new explanation of the autopoesis (self-organization) of innovative and creative societal forms. Just as new species evolve under the impact of mutations, new societal forms can originate from unforeseen random events.

Today, it is an enormous advantage for the explanation of societal evolution that we possess two formulations of optimal conflict solutions. One is the psychological form: we optimize the expected utility and modify it by risk. The other, stemming from the matrix form, is Maynard Smith's dynamic definition of optimal solutions by relatively or evolutionarily stable societal forms. The matrix formulation expresses the individual's psychologically founded maximization of its individual and societal stability and security, and not any longer the maximization of its egotistic utility. Both formulations show how, in the course of societal evolution, relatively stable societal solutions become identical with an optimal, advantageous adaptation of the individual's solutions of conflicts to changes and random events in his environment. Sociologically, adaptation can be explained by a slow and gradual modification of individual or group behavior or by sudden (punctual) adjustments of social life to a changing cultural or other external environment. Selection, again, works in any evolutionary conflict-solving process when certain individuals favor optimal solutions in the struggle to achieve individual and collective stability in case they are disturbed by random events. It expresses the general tendency of individuals to prefer safe solutions to unsafe ones. Thus, the dynamic aspect of societal evolution can be characterized far better by the attractor of societal stability and the successful perpetuation of evolutionarily stable solutions, than solely by the individual maximation of utility. If we add to this Axelrod's cogent explanation of the autopoesis of altruistic or cooperative conflict solutions (Axelrod 1984: 88f, 145f; Leinfellner 1986: 141-147), we understand why symbiosis and social cooperation and association - for example, as teamwork - became the leading attractors of societal, cultural, and ethical evolution.

8. FROM DYNAMIZATION TO EVOLUTIONARY SCIENCES

This article shows that three recent trends shook not only traditional static game theory, but also spearheaded the "dynamization" in most of the social sciences. Firstly, the "dynamization" of the traditional utility, game, and decision theory simply means that evaluations, decisions, and conflicts are in most cases serially connected, temporally ordered successions of conflicts and of their solutions. "Dynamic" means that societal life is a constant, never-ending struggle to solve emerging conflicts. These evolutionary conflicts have to be solved and terminated. Open unsolved conflicts lead to chaos and decay. Therefore, evolutionary game theory and the stochastic theory of evolution deal exclusively with optimal solutions of serial conflicts. During millennia of societal conflict solving, innumerable rules, ordered in specific batches, have been created, accumulated, and stored, more and more democratic welfare principles and rules among them. The preliminary result of this historical evolution are our democratic welfare societies and their ever-increasing rules. Secondly, the social sciences deal with the creative and autopoietic character of evolutionary serial conflict solutions and their output of new rules and solutions. Thirdly, the new psychological roots of utility theory and of evolutionary game theory made possible the new psychological foundation of all dynamic social sciences, as far as they deal with serial conflicts and their solutions.

Above all, dynamization means - on the theoretical level - the change of the traditionally linear and static sciences to evolutionary, non-linear ones. The solution of a serial conflict cannot remain forever in an equilibrium or stable state, since any static equilibrium would be the death of evolution. The trajectories of societal evolution approach only relatively stable, that is, dynamic, equilibria which build up and stay relatively stable for some time only; sooner or later their stability will be disturbed by environmental changes and/or inner random events and may break down. The result of advantageous selective pressure shows up in evolutionary models as relatively stable solutions. Non-advantageous pressure will show up in unstable solutions. The approximate solutions and the cyclic solutions belong to this class. For example, the famous business cycles are, as their name indicates, cyclic solutions. For all these stable or also quasi-stable solutions a common description is being used: the attractor of an evolutionary process is independent from the initial conditions (Eigen & Schuster 1979: 32 ff). It is understandable why, from the seventies onwards, biology has adopted these new concepts and methods of evolutionary game theory. But nobody expected their successful transfer to, and similar use in, evolutionary cosmology, chemistry, and physics (Leinfellner 1983: 135-147). Finally, in the last years, the new theory of evolution gained a footing in economics, the social, and the political sciences.

9. THE TWO PSYCHOLOGICAL FUNDAMENTALS OF TODAY'S THEORY OF EVOLUTION

Modern risk theory (Allais, Munier) and Maynard Smith's "relative stability theory" (Maynard Smith 1982: 14ff) are the two fundamentals of the stochastic theory of evolution. One deals with psychic utility and risk, the other with the concept of societal stability. Utility theory explains the evolutionary dynamic by the changing psychic preferences and expected values; risk theory takes into account the sudden changes or random events. Values are obtained from psychic preferences, liking or disliking, pleasure or displeasure. Evaluations of the solutions of economic conflicts can be measured by individual values or utilities (money, financial rewards, risks) either on an ordinal or a cardinal scale. Values can be compared and measured even on multidimensional scales. Moreover, values can be combined with probabilities; this describes the internal psychological and/or genetic roots of how we begin to solve conflicts. Even though evaluation under uncertainty and risk functions unconsciously, we can render them conscious; and we can reconstruct them in the models of utility and game theory under risk and uncertainty. Conflict solving has been - wrongly - identified with rational behavior; but modern utility theory does not stop here. Conflict solving is based on decision rules, not on deductive principles. Deductive, rational decision-making, especially the maximization of values (utility) or even expected values (utility), is not the sole criterion for conflict solutions. According to Allais, the highest criterion is risk; on the maximization of expected utility Allais' model imposes the internal criterion of risk (Allais 1979: 6, 67-73, 480). Thus, we modify the maximization of utility by our risk attitude; risk averting and risk loving are rooted in our psychological and genetic makeup. Animals and humans do not maximize the theoretical "survival of the species"; instead, each individual follows its genetically rooted risk behavior. Therefore, evolution is guided by the individual's preferences, utility, and risk (Grunert 1989: 391).

10. THE FREQUENCY INTERPRETATION OF SOCIETAL EVOLUTION AND THE ROLE OF EVOLUTIONARY STABILITY

In biological theories, for example, in the Darwinian theory of evolution, fitness provides a natural and genuinely one-dimensional ordinal scale. Albeit the concept of human preferences and evaluations is difficult to apply to the evolution of animal contests or conflicts, fitness can be expressed easily by the highest and most stable frequency of a certain trait, a custom, a societal form, or of how animals optimally solve a conflict. Thus, fitness as the inborn strategy of survival of the survivor loses its Darwinian - and tautological - character when it is replaced by risk-modulated expected utility, and when one introduces the concept of evolutionary stability as the attractor of evolutionary paths (Maynard Smith 1982: 19).

Luckily, evolutionary stability is an extension of the concept of static equilibrium to serial solutions of conflicts; and it could be applied successfully within the biological theory of evolution. As early as 1951, Nash showed that every non-cooperative conflict has at least one equilibrium strategy as optimal solution; cooperative game theory again rests completely on the concept of equilibrium as the solution of conflicts. In animal conflicts, preferences, liking, disliking, utilities, values, and risk behavior are difficult to observe; moreover, in most cases they are the expression of an unconscious, automatically and genetically functioning adaptive, risk-guided mechanism. Therefore, evolutionary stability is a perfect criterion for solutions of serial evolutionary animal conflicts. The experimental situation has changed, too. Recent research by Damasio, Battalio and Hursh provides clear experimental evidence that the basic principles of preferential, evaluative, and risk behavior can be observed when animals solve serial conflicts (Hursh 1989: 393-396, 406; Damasio 1994: 83 f, 187 ff, 197 ff). Today, we can measure an animal's preferences, values, and risk behavior in the same way as we measure human values.

But in spite of the fact that the stochastic theory of evolution explains both the animal's and the human's genetic and societal evolution as the way of how they solve serial, particularly intra- and intergenerational, conflicts under uncertainty and risk, they differ with respect to the empirical interpretations. Both forms of evolutionary conflict solution occur within populations or proto-societies; but their past experiences (histories) are stored in different ways. In animal societies, they are stored mainly in the preserving and protecting genetic memory which is inherited and then evoked by imitation and learning without conscious teaching; in human societies, besides being stored in the genetic memory, they are stored mainly in the individual brain memory and in the collective, societal, and linguistic memory. Memories transmit information, rules, and experiences of how to solve conflicts by learning, teaching and language (Tulving 1972: 9, 24). Human societal evolution is the modification of societal forms; these forms are stored and handed down consciously by teaching and learning within or across generations. Societal forms are: customs, behavior, institutions, mentifacts, and artifacts (artistic and technological). Societal conflicts occur between existing or changing societal forms. In contrast to animal evolution, cultural evolution does not necessarily entail any form of genetic change. But when co-evolution begins to operate, then the two lines of evolution, the biological and the socio-cultural, are, over long periods of time, coupled together in an interactive manner, according to Wilson and Lumsden (Lumsden & Wilson 1981: 11, 237-305). Gencultural co-evolution means that important life-preserving instructions which enhance a society's stability are stored in our cultural heritage; over a long period of time, they may become encoded in our genes, too.

The concept of relative or evolutionary stability expresses the social fact that human evolution depends far more on a society's stability than on Darwinian survival. For example, when a certain trait or custom increases its frequency within a society more than any other trait or custom, we speak of its successful evolution. There are many good reasons to expect that serial solutions of conflicts between customs, etc. evolve in the course of evolution until they reach a relative or evolutionary stability which creates relatively stable states. In the Darwinian theory of evolution, individual utility, preferences, and risk play a minor role; but in today's reformulations of societal (economic, political, social, and cultural) evolution the reverse holds. Preferences, values, utility, and risk are the driving psychological forces of any societal evolution (Damasio 1994: 186f, 197 ff).

11. THE REALIZING FUNCTION OF THE SOCIAL SCIENCES AND THE URCONFLICT OF SOCIETIES

Realizing is the act of converting an - imagined or computed - solution (optimal strategy, plan, or model) into deeds, actions, reality. It is a difficult task. For example, democratic welfare principles (human rights) can only be introduced and realized successfully if a form of market economy has already become part and parcel of a nation's makeup, not vice versa, as the examples of China and certain African countries prove. Finally, risk taking is the readiness or chance to encounter harm, losses, or winnings, caused by random events, when we realize a certain solution. Risk indicates the individual's willingness to being exposed to a reality which is haunted by randomness in terms of future losses, advantages, etc. In the stochastic theory of evolution, one begins with preference theory and utility and risk theory; one ends with evolutionary game theory. The purpose of this stochastic theory is to serve as a new tool for optimally solving evolutionary societal - economic, social, political, cultural, and even ethical - conflicts under uncertainty and risk, in other words: always to be ready to terminate open conflicts and crises. Uncertainty means that we are always under the constraints of unforeseeable changes which are caused by random events; risk means that we are under the controlling and steering influence of our psychological and genetically inbuilt risk "factor", vis à vis random events. In brief, the purpose of the stochastic theory of evolution consists in explaining how our intelligent risk behavior masters "uninsurable" risks or randomness (Damasio 1994: 173).

Today's economy and societal life is haunted by suddenly appearing random events, changes, and crises, and by the ups of prosperity and the downs of recessions. A new theory which takes into account our fears and hopes - fears to lose our individual and societal stability, security, and hopes that stability will increase again - comes just at the right moment (Hagen 1994). The stochastic theory of evolution delineates in new terms and formulations the urconflict of all humans and their societies: how to survive. In more theoretical terms, we search for relatively stable solutions when random events cause sudden and disturbing changes. The readiness of human beings to incessantly solve this urconflict against randomness rests - since the rise of homo sapiens - on a sequence of intelligent steps. The first is a practical one. We observe the regularly occurring events and store them in the form of rulelike laws ($rule_1$) in our memory; and we carefully watch the random events and their consequences in our society and our environment. The second step is to evaluate both, the reoccurring, causally linked events, things, customs, artifacts and mentifacts, and the random events; their evaluations are, in turn, being stored in our subjective and collective memory. The third step rests on our intelligent, Promethean ability to cope with our future by combining and recombining evaluations with probabilities of random events and to draw inferences of how to solve optimally new and old conflicts with the help of rules^1 and socially accepted, democratic rules^2. Rules^2 are conflict-solving default rules. Only this form of intelligent activity takes into account the uncertainty of information vis à vis random events; and it alone is guided by the controlling modification (alteration) of expected solutions of conflicts by the individual's and the group's risk attitude or risk "factor". In a last step, our intelligence aims at the realization of imagined solutions of conflicts (plans, strategies, prospects). Once these solutions are realized, we learn from their empirical confirmations. Finally, we store these experiences away for future use.

12. THE ROLE OF EVOLUTION WITHIN DEMOCRATIC SOCIETIES

This article restricts societal evolution to democratic welfare societies. Only democratic societies offer an ideal leeway for fair conflict solutions. The main reason is that such societies store their fair and just juridical rules, laws, and principles for batch processing. A batch is a unit; it contains a specific group of data and rules for solving special societal conflicts. Democracies even permit the improvement and adaptation of these batch rules as well as of their constitution if it is necessary. Juridical, moral, and ethical decision rules which normally govern the public decisions of individuals and groups within democratic states can easily be incorporated into evolutionary solutions of societal conflicts. Democratic rules for solving public conflicts function in a "hereditary" fashion similar to the encoded behavioral instructions in our genome; i.e, they are stored, transmitted, and available whenever we need them. Understanding the code of human and societal life certainly will be one of the greatest revolutions in scientific knowledge. Finally, democracies are rule preserving and innovative. If the batch of inherited data and rules does not help us to find a solution, we may create new rules if the community agrees.

Thus, democratic societies are a sheltered haven for the storage of optimal decision rules. The stochastic models of evolutionary game theory, then, are not only reconstructions of intelligent serial conflict solutions within democratic societies; they also use the fair rules and principles of western democratic societies. This began with the incorporation of democratic rules into game theory by Morgenstern and Neumann, by Luce and Raiffa, and into collective choice theory and into ethical decision theory by Arrow, Sen, Harsanyi, Leinfellner, and many others. Innovative rules and new solutions of evolutionary conflicts by the individuals within mini-societies (committees) and societies can push forward optimally the evolution of democratic societies. Especially in democracies, the individuals are not just passive onlookers of their society's evolution but can function as actors and agents. In the long - but not too long - run, only primary (direct) and secondary (indirect) democracies guarantee socially fair solutions of economic, political, and social conflicts. These solutions originate directly and indirectly from the acting, deciding, evaluating, and risk-taking individuals.

Today, the stochastic theory of evolution permits us to separate the societal evolutionary paths (the economic, social, political, and cultural evolution) neatly from the biological ones. For example, the Darwinian evolution was dominated by the idea of strict competition; the democratic, societal evolution is mainly guided by cooperation (Axelrod 1984; Lumsden & Wilson 1981; Leinfellner 1989). This is another reason why we restrict our analysis of the evolution of societal forms to the present western democracies. Historically speaking, they are the last stage of an ongoing societal evolution which has already lasted for millennia. In contrast to the Darwinian theory, the societal evolution leads to the storage even of individually acquired properties and democratic decision rules, provided they contribute to optimal societal solutions; this resembles Lamarck's view. Another, mostly unnoticed difference between societal and biological evolution is that the Darwinian maximization of survival tells us only the egotistical half of the story. According to the stochastic general theory of evolution, biological survival is practically always guided by the individual's innate risk attitude. This fact can be used to support the foundation of economics on the psychologically and genetically anchored, individual and collective risk behavior of human beings in their democratic societies (Allais 1953, 1978: 37-160; Lopes 1988: 405-416).

Another difference is the leading role of cooperation and teamwork in all forms of societal evolution. Its basis is Axelrod's proof that only within democratic societies do cooperative solutions of conflicts arise automatically and autopoietically, and even from egoistic behavior (Axelrod 1984: 1-88, Leinfellner 1988: 244-248). This result allows us to put competitive or cooperative conflict solutions, societal change, dialectic development, progress, chaotic phases, and revolutions under one roof. They are facets of evolution; we are able to subsume them all under societal evolution. Of course, evolution cannot proceed in a competitive and a cooperative manner, in a dialectic and a chaotic way at the same time; but it will pass at different times successively from one phase to the other.

13. EVOLUTIONARY TRAJECTORIES

It makes no sense to speak about evolution as such; it manifests itself only as specific evolutionary trajectories. As already mentioned, evolutionary paths or trajectories can be conceived as composed serial, temporally ordered solutions of one or more conflicts; for example, when individuals with their experience and their tradited knowledge favor a new societal form and reject an older one. Only these evolutionary serial conflicts enable the individuals to learn. Learning is our best strategy, the science, plan, or technique for achieving that evolution is not the exception and extinction not the rule. Since conflicts consist of open alternatives, possible solutions, bifurcations between possible acts, moves, strategies, projects, and prospects (probable alternatives), it is clear that open or undecided conflicts are dangerous crises. Only optimal solutions successfully terminate conflicts. The theory of evolutionary games and the stochastic theory of evolution can supply us with optimal solutions and terminations of evolutionary conflicts; but both cannot hinder the outbreak of a new conflict. Therefore, to realize an optimal solution means in most cases only to initiate or open up a new evolutionary path or to reuse an old, already proven one. But we have no guarantee that the final goal of an optimal solution can be totally realized. Realizing is restricted only to initiating or channeling the direction of an evolutionary path for a foreseeable time. The realization of any optimal solution is a provisional one: we hope that no unforeseen random events will disturb the new path, either immediately or later on.

14. THE STOCHASTIC THEORY OF EVOLUTION

Albeit the analogy to biological evolution is striking, any identification of the societal with the biological evolution is misleading, mainly because evolutionary game theory and the stochastic theory of evolution have changed the basic ideas of Darwinian theory, weighted them differently, and incorporated many other, equally potent democratic social principles and rules in a more precise and mathematically usable form. For instance, the concept of socio-cultural evolution changed the merely biological "automatic" of selection to an instrumental one. The new concept of memory storage and traditing of instructions (rules[2]) changed the concept of inheritance: instead of biological inheritance, we have learning and cultural tradition as the regulating factors. Individual and group survival are being reformulated as the maximization of stability and security. Finally, it was possible to prove the superiority of cooperation over competition within the evolution of democratic societies. All these concepts were reformulated in a mathematically applicable form. Today, as a consequence, the stochastic theory of evolution comprises biological, economic, political, societal, cultural, and even ethical evolution as special cases; but it provides for each a different empirical interpretation.

In our context, "stochastic" or "general" means that evolutionary serial solutions of conflicts do not depend any longer solely on selection, adaptation, and survival, and on societal and environmental random events but also on active interferences (successful or unsuccessful decision making) in our cultural evolution by individuals, groups, and society as a whole. In short, serial decision making is a multidimensional decision process depending on many external and internal, hierarchically ordered criteria. The weight of these criteria changes from case to case. For example, in competitive and cooperative solutions, we are more dependent on the other members; in solutions against random events, more on randomness and on our countermeasures. "Hierarchical" means that the risk factor is the highest criterion; it modifies even our expected evaluations - under the proviso that societies and their environments depend far more on random events and our risk attitude than we have previously assumed.

15. RELATIVE STABILITY AND BREAKDOWNS IN THE COURSE OF EVOLUTION

Even when we solve our evolutionary societal conflicts optimally, we can secure the stability of a preferred societal form only in a relative sense, for a certain time. "Relative" means that in evolution nothing is permanent and forever. Even optimal solutions last only until the next disturbances by bigger random events or until our interferences and voluntarily induced changes stop or shatter them. There is one exception: the storage of instructions in the biological genome and in our collective cultural memory seems to last forever or at least as long as life on earth and human societies exist. All such past instructions are stored away and batches of instructions are available from the two memory storage systems if we need them. The storage and availability of past experiences and decision rules plays the same role in any kind of societal evolution as the storage and function of genetic instructions in our genes; but their role and functions have quite different empirical interpretations.

Inability to solve our conflicts because no past rules exist, or because we don't invent new ones, causes a breakdown of our societal stability. In this case, it may happen that the never-ending biological evolution takes over the societal evolution. If we are lucky, this may lead to a new optimal solution. But if we have no such luck, society may yield to genetically rooted, merely biological instructions; it will slide into archaic solutions, such as strictly competitive racial fights between groups, into tribal warfare, chaotic phases, into the kind of debacle which we see today in many countries lacking democracy.

Any identification of selection and adaptation in biological evolution with selection and adaptation in societal evolution is misleading. Yet, an analogy does exist. In both "strands" of evolution, unexpected random events play, from time to time, the role of guardian angels which cause innovative creations, such as new species and new democratic societal forms, ecologically oriented markets being an example of the latter; but from time to time, they may have disadvantageous or even deadly consequences, like outbreaks of racist violence.

That is exactly why St. Gould regards biological evolution as a lottery in which we have been the winners - until today, at least. Yet there is a difference between a commercial and an evolutionary lottery. In the former, random distribution begins with each trial anew; its randomness is blind and cannot be influenced. In any evolutionary lottery, randomness can be partially influenced: to a higher degree by human beings, to a lesser degree by animals, as we will show in this article. For example, we may influence the likelihood of criminal actions by laws, or prevent the flooding of a riverbank by channeling.

Quite contrary to biological adaptation, societal adaptation, i.e., the process whereby we adjust ourselves to new conditions, to environmental and random events, is taken over more and more by our conscious, scientific and technological realizations of intended solutions of evolutionary conflicts.

One can easily prove that serial optimal solutions are adaptational processes which may, for instance after an exogenous shock, lead to a new equilibrium; non-optimal solutions lead away from equilibrium points. Each realized solution, then, is an interference in the "ongoing" societal and sometimes even in the biological evolution. In spite of the fact that any societal realization interferes with the course of evolution only in a partial (statistical) sense, it tells us more about our next future than predictions. Whatever its character, it induces a new direction of the evolutionary societal trajectory. Even the realizations of optimal solutions channel the evolution of society only for a certain period of time. If we are lucky, realizations of evolutionary solutions influence the direction of a newly created evolutionary path for a period of time which is longer than planned. The basic idea in the background is that any interference with evolutionary trajectories amounts to a game between us and evolution. Therefore, dynamic stochastic models are stochastic reconstructions of possible optimal and relevant evolutionary solutions; as a byproduct, they often create new rules which may help us to optimally solve societal conflicts. "Optimal" indicates neither the maximation of expected utility nor the best solution for me or my nation but the solution with the best chance to realize a relatively stable state of my society or mini-society, my family, and the like (Mesterton-Gibbons 1992: 45-83; Leinfellner 1984: 252-256).

16. DEFAULT RULES

The historical conceptual change from static equilibria to dynamically shifting equilibria had another consequence. It caused a replacement of all traditional hypotheses and so-called social laws by default rules (Holland 1986: 153f, 282). Firstly, statistical, or default, rules (Holland 1986: 22, 68-

97) are inductive, not deductive (E. Leinfellner 1992: 363, 1994: 88 ff); this is typical for all evolutionary processes. "Default" is a new way of expressing the old adage "No rule without exception" - i.e., exceptions must enter into rules to bestow on them an inductive-inferential function. Secondly, default rules are prescriptive, not normative. Batches of them are instructions for realizing or executing patterns, customs, and societal forms, and for solving conflicts; these instructions are accepted by individuals and groups. Thirdly, default rules comprise not only behavioral, structural, and decision rules, maximizing rules, the Bayes rule, and rules of inference but also democratic and ethical rules and principles, for example, the rules of Pareto-optimality, of equality (A. Sen), and of justice à la Rawls. Bamberg and Coenenberg (Bamberg & Coenenberg 1991: 89) classified rules along the same lines. According to Holland & al., all these rules are the methodological, "prescriptive" backbone of all inductive inferencing and of non-linear conflict solving.

17. THE TRANSITION OF SOCIETAL DEVELOPMENT TO AN EVOLUTIONARY PROCESS

In the seventies, at a time when evolutionary game theory began to solve serial conflicts, the incorporation of new evolutionary concepts triggered a snowball effect, the development of the stochastic theory of evolution. It began with the replacement of the traditional utility theory "without risk" by the new utility theory under uncertainty and risk. From now on, this new utility theory served as the psychological fundament of evolutionary game theory. Risk turned out as the secret attractor of evolution. It tends to terminate conflicts as soon as possible and to avoid unnecessary losses. For example, the traditional computation of competitive serial solutions by maximizing the expected utility is now to be constrained by risk factors. Repetitions of serial solutions accumulate rules and permit learning and the improvement of past successful or unsuccessful solutions. Both rules and solutions are stored in the individual or collective memory, our history. No player would be so stupid as not to learn from his or her previous experiences. Knowledge of history acquires the character and role of a serial supergame which includes historical and also present games as subgames (Friedman 1986: 80, 72). Evolutionary game theory permits not only solutions of competitive and cooperative conflicts but also solutions against and "with" random events; like mutations, these random events may cause losses or winnings. Here the resemblance to the double role of good and bad mutations in biological evolution is obvious. Aside from learning, evolution is an autopoietic process; it is the only known process which has such a high output of creative innovations, new solutions, and new rules.

Today, the general evolution equations, the genetic algorithms, and the models of evolutionary game theory and its matrix method are the new "software" for computing, simulating, and even creating possible evolutionary societal trajectories by computers. Lumsden and Wilson's replacement of the hereditary units of the biological evolution, the genes, by culturgens, permitted reconstructions and simulations of societal evolutionary trajectories. High frequencies of culturgens indicate their relative stability. Phenotypically, culturgens show up as artifacts, behaviors, institutions, mentifacts (i.e., mental constructions, ideas), etc. They are culturally transmitted (inherited) by learning within or across generations (Lumsden & Wilson 1981: 368). For example, free market behavior is shared by the majority of western people and transmitted by learning within and across generations.

These examples show how the function of selection, adaptation, storage, and transmission can be incorporated into the stochastic theory of evolution. Within this theory, their formal function stays the same; but their empirical interpretation differs in quality, degree, form, and material from any biological function. Analogically, the new theoretical framework of the general theory of evolution with its stochastic methods of computing possible solutions of serial conflicts can be used to explain the different forms of societal, economic, political, social, cultural, and ethical evolution.

As already discussed, all optimal solutions express mathematically and empirically the most important evolutionary property of societal forms: their relative stability, called "evolutionary stability" by Maynard Smith (Maynard Smith 1982: 16, 204).

The good news about stability is: what stays stable cannot be easily overthrown by small and everyday disturbances, random events, etc. An inveterate Darwinian would say: if anything which is stable will survive, why not work with survival in the first place? A defender of the stochastic theory of evolution will prefer the concept of relative stability; this concept can be quantified and mathematized, and therefore used in his/her theory. The maintenance of relative stability also justifies our interference in the course of evolution; it has to be strong enough to effect the desired goal. If corrections are repeated often enough, the conflict solutions may approach a certain relatively stable state or point. Such a series may be a long-lived evolutionary trajectory.

The bad news is: any stable phase in the course of evolution is only relatively stable. Unforeseeable disturbances, any random event above a certain threshold may upset stability. Such a trajectory may then cease to be stable and fall victim to - often chaotic - periods.

18. REALIZATIONS OF SOLUTIONS

The fact that our "artificial" realizations of solutions can exercise a far greater influence on the future evolution of our society than on biological evolution makes a big difference. It is true, we cannot realize the planned course of an evolutionary trajectory with 100% effectiveness; in this case, our predictions would be deterministic. Throughout our history, we were sometimes lucky and sometimes unlucky when we interfered in the biological and societal evolution. In former times, we didn't know too much about evolution; even Popper regarded our interference more as a trial-and-error procedure than as an effective realization of optimal solutions. The crux lies in the fact that even today we cannot be sure that unpredictable random events will not interfere in the realization or in the successfully started (channeled) trajectory or course of our society's evolution. It seems that in times of frequent interferences by random events, evolution gets automatically the upper hand and forces us to share our influence to a great extent with random events. In times of lesser interference by random events, our influence on evolution gains more weight. The higher the randomness and its intensity, the less influence remains for us, vice versa. Channeling can be compared to throwing switches for shunting railroad cars in a railroad yard: some are being thrown by ourselves; others by randomness.

It goes without saying that societal evolution can neither be explained fully in terms of utility and purpose, nor teleologically in terms of plans or final causes. The real reason lies in the fact that any interference by us or by random events will disturb any linear process and change it to a non-linear one. But in the long run, we may have greater chances to influence our societal evolution by iterating realizations of our optimal solutions. The iteration of realizations of solutions channels evolution like a man-made riverbed channels a river's flow of water. We may initiate and direct the flow of the water; but we can never be sure that the water will run forever in its artificially made bed. All we can do is to be ready to repair the channel if this should be necessary.

19. THE PARADIGMS OF THE EVOLUTIONARY AND THE COMMERCIAL LOTTERY

Besides the paradigm of the game, another paradigm exists which is closely connected with the former. It is not so well known; but it has been used successfully by all utility and value theorists to measure values. It is the lottery paradigm, a primitive Kuhnian paradigm of our dependency on uncertainty and risk. With or without Kuhn, the Greek word "paradigm" means "model" anyway. The model of the commercial lottery explains the distribution of prizes. Chances are sold in the form of numbered tickets; the winning tickets are selected by lot or chance. It is clear why the paradigm of the evolutionary lottery turned out to be the protomodel of how we cope with random events with the aid of our risk attitude; or, expressed more generally, it gave us an explanation of how we solve conflicts in which the likelihood of success is uncertain and our risk guides us to find an optimal solution between expected losses and winnings. That is the reason why some people regard even life and marriage as a lottery.

Paradoxically, one side of this paradigm, which expresses our most genuine evaluation of uncertain states of our world or our society, has been used by Neumann and Morgenstern to measure values quantitatively. In the traditional preference and utility theory, it has been "theorized" and used to determine quantitatively our dependency on random events. Interestingly, the early American school of utility theory neglected the role of risk but regarded its own utility theory as the cornerstone of today's economics. Following Allais' pioneering research on risk, the other side of the lottery model, its risk dependency, attracted more and more attention. It is no wonder, then, that it began to play such an eminent role in the stochastic theory of evolution.

In utility and value theory, the uncertainty caused by random events is expressed by probabilities x_i, allotted to the values of conflicting and uncertain alternatives, to actions, strategies, and trajectories $a_1, ..., a_n$. Like in all lotteries, the model permits to compute the total value u of all expected alternative values $u(L) = x_1a_1, ..., x_i,a_i$ or simply, $u(L) = x_1,a_1 + ... + x_i,a_i$. It deals not only with simple lotteries but also, and mainly, with compound lotteries or with lotteries of lotteries; this throws a new light on the complexity of solving evolutionary conflicts. This aspect of the model also demonstrates how we combine in our imagination simple lotteries to compound ones and how we can compare and compute their expected values in our mind.

Here we will analyze the psychological, mental, and stochastic role of lotteries and compound lotteries in evolutionary conflict solutions. The forms of compound lotteries represent typically cognitive patterns: they have led to the mathematical reconstruction of prospects of future evolutionary alternatives or trajectories from which we select the optimal ones. Now, if we replace normal alternatives x_ia_i in simple lotteries by sublotteries x_i,L_i, they represent in our imagination, in our mind the different future trajectories of evolution and the possible interference of random events. In brief; in our mind, we see our future as expected future evolutionary paths. In the general theory of evolution, sublotteries represent stochastically the expected values of different evolutionary paths which lead into the future. This interpretation also shows, firstly, the psychological aspect of how we see in our imagination our and our society's uncertain future evolution before we attempt to realize a solution of a typical evolutionary conflict.

Secondly, this psychological interpretation also shows the striking similarity of our psychological attitude before drawing a ticket in a commercial lottery when we ponder about the different future possible losses and winnings.

This internal aspect, as depicted by the lottery paradigm, belongs to our stochastic world view; it cannot be explained away by logical, deductive, rational methods, or by logically possible future worlds. Through the lottery paradigm, we see mentally our (not too far) future evolution, unfolding in many evolutionary paths. For the first time, this evolutionary model, the lottery model, takes seriously into account the randomness of our world and society. This mental processing is a parallelly distributed intelligent procedure, as described by Rumelhard and McClelland (1986). We look into our uncertain future in a parallel, not in a serial, way; at the same time and in a parallel fashion, we comprehend the possible courses of evolutionary trajectories. "Parallel" is used in the framework of Rummelhard's concept of cognitive apprehension: we oversee mentally and as with one glance many conflicting sublotteries and pick up the one which is optimal.

There is another difference between commercial and evolutionary lotteries. In an evolutionary lottery, we are not only in the situation of someone who plays constantly in different lotteries and can choose between them; but we can also change and create new lotteries by ourselves, for example, change the random distributions in favor of ourselves.

The psychological roots of taking an "evolutionary" lottery ticket depend on my ranking of random prospects, on the valuation of my initial capital, on the expected gains, and finally on my risk attitude (or my hopes, fears, and expectations). Thus, taking an evolutionary lottery ticket illustrates psychologically the same situation as trying to solve a future evolutionary conflict between random prospects or evolutionary trajectories.

The difference between commercial and evolutionary lotteries explains why we, in our society, are not only passive onlookers of evolutionary lotteries. In this kind of lottery, we are also active agents. For example, we can create, organize, and realize evolutionary lotteries and, using technical means, provide that random events have no catastrophic effects. As "owners" of lotteries, we dispose, to a certain degree, of our societal evolution in the same sense as we manage lotteries and adapt them to democratic statute laws. Thus, the evolutionary lottery paradigm describes not only our inner psychological view but also our dependency on random events and our possible protection from future devastating random events. It reveals the urconflict of all life and its never-ending struggle against sequences of random events. Our individual and societal future depends on how we solve this urconflict. There is no doubt that the evolutionary lottery models describe the urconflict with randomness and our mounting scientific and technical influence on our evolution.

After the foundering of the mechanistic world view and its deterministic predictions and prognoses, the theoretical reconstruction of the evolutionary lottery paradigm within the stochastic theory of evolution offers a new explanation of our cosmic, our environmental, and our societal evolution, including the attending random events. Our technical interference in our environment, our intervention in our society, our culture, and even in our planetary and, according to Tippler, in our cosmic evolution must be compared to playing in evolutionary lotteries; it is the most complex and most difficult conflict faced by mankind. The question is: can we expect to solve this conflict optimally? If we can't do it today, we have a good chance to find better solutions tomorrow, provided that our scientific knowledge will increase. The lottery paradigm and its reconstruction show us a heuristic and stimulating method of how we can use our intelligent stochastic abilities which we have practiced automatically since times immemorial. It shows how we can push back and minimize the influence of randomness on our evolution. So far this has been a good strategy; otherwise we would not have survived.

A main strategy appears: research into our intelligent abilities depends quite generally on the growth of our knowledge of evolutionary processes. According to Götschl, research in evolution will certainly be the most promising strategy to solve the urconflict with the blind random events. The more we know, the better our winning chances in this lottery against randomness. Only scientific knowledge will prevent evolution from being the exception and extinction the rule.

20. DEFINITION OF RANDOM EVENTS, NATURAL LAWS, REGULARITIES, AND RULES

For Kant, philosophy had to answer three questions: What can I know, What shall I do, What can we hope or expect? Traditional philosophers and scientists assumed that the classical laws of physics, presumed to be universally valid, would give us an answer to the first question. The default rules of our societies, including the ethical ones, which take into account exceptions and deviations, would answer the second question. Combined with conflict-solving models, they tell us what we can expect from our future in spite of unpredictable random events.

Since the fifties, faith in the unlimited validity of natural laws has slowly vanished from the physics books. The law of the preservation of energy seems to be the only one left over. It alone can be applied without exceptions until today. J. A. Wheeler goes so far as to state: "There is no law, except the law that there exists no law.". As a consequence, today's natural laws resemble more statistical "all"-statements or default rules rather than Popper's universal (universally quantified) "all"-statements. It makes no difference that rules are prescriptive and natural laws descriptive and that both differ with respect to their field of application. Today, they have a lot in common. They are both statistical, highly confirmed statements. According to E. Noether, they express that there is always a magnitude which stays invariant or relatively stable (Genz 1990: 31, 131, 441). This concurs with the fact that optimal decision rules also keep one magnitude invariant: the relative stability of societal forms. From Noether's point of view, social sciences search for stable societal solutions rather than for truth. Since they are relatively stable, societal forms possess the highest degree of order. Any optimal solution of evolutionary conflicts is a search for relatively stable or ordered societal forms. This is exactly the way in which evolutionary solutions and realizations create societal order. In case this order is disturbed by random events, we can recreate a new order, and so forth. It is a paradox of evolution that out of disturbances, randomness, and chaotic states order can originate.

If we try to answer this paradox, we come to the big question, unanswered till today: what is a random event? Traditionally, random events have been defined negatively as offsprings or causes of disorder, chaos. Even more paradoxical explanations exist. Random events happen in absence of any evident cause; they occur by chance rather than by design. We already know that chance has a Janus face: one face symbolizes winning and the other the losing hazards. Interestingly, in evolutionary societal conflicts, random events share these "blind" properties with biological mutations and randomness: in both cases, chance or hazard is "synonymous" with risk or gambling, or, finally, with a ticket in a lottery.

We know that random events can be generated naturally and artificially; according to Allais, there is no difference between these two kinds of randomness (Leinfellner 1989b). It seems that series of random events can be generated when we replace natural random generators by artificial ones. Some generators use natural microscopic physical random processes, such as a thermal electron current in a solid-state diode or a radioactive decay which transforms its generated random "noise" into spaced strings of randomly alternating, binary pulses or "bits", like 0, 1, 0, 1, 1, 1, 0, 0, 0, ... Astonishingly enough, newest research by B. Arthur proves that democratic societies can be regarded as natural random generators!

Our deeper understanding of randomness advanced when we abandoned the classical Einsteinian one-one causality where one effect always follows one cause, in favor of statistical causality as used in quantum theory (Leinfellner 1981: 253-258; 1989: 199f). Statistical causality again is based on Bell's theorem. According to Bell, in complex, hierarchical systems all events are directly or indirectly causally interdependent. For example, we cannot separate a single event from a system without changing the whole system. Therefore, statistical causality assumes that an event can have not only one, as Einstein assumed, but also many or innumerable partial causes;

each of them possesses a certain "causal" weight which can be observed and expressed by fluctuating probabilities, as Heisenberg and the quantum physicists assume (Thorne 1994: 430). Such a fluctuating many-one causation is called a causal fork; it can be found in most non-linear systems and in game-theoretical trees.

According to B. Arthur, societies are interdependent, complex holistic systems. They contain causal forks and are, therefore, natural random generators. They and their environment are governed by statistical causality; therefore, they can generate random events on their own. We can now define a random event as the generated effect of innumerable fluctuating partial causes. But that is not all. In the course of time, the partial causes of dynamic systems change their weights; their oscillation will lead to an unexpected fluctuation of their resulting effect, which we call random event (Leinfellner 1987: 199ff). Whether we like it or not, social systems are random generators and generate random events. The more complex a system, the more random events it will generate. Important complex subsystems are our brain with its neuronal random firing, democratic societies, our environment, the cosmos, provided that they interact evolutionarily.

For example, complex societies, such as ours, increase their frequency of random events during recessions. According to McConnell, recessions are caused by random events. The accumulation of disturbing negative accidents has a disastrous effect on business cycles. The ups, or prosperity, create unexpected random or windfall profits, the random events of each depression, unexpected and uninsurable losses (McConnell 1966: 585).

One interesting randomizing societal event is the free will of individuals in democratic societies. For example, arbitrary, unpredictable choices, unpredictable assents to one of the alternatives of a conflict as the result of the individuals' free will may be dangerous because of their possible aggregation. For example, consumer tastes may aggregate; risk attitudes may change without any reasons. If, out of the blue, they aggregate, they may influence drastically the course of societal life, of the economy, or of elections. It is an inconvenient new fact that modern mass societies tend to temporary chaotic phases because of their individual freedom of choice, the interdependence of their elements, and their complexity, and because aggregation is facilitated to a high degree by the media. This may enforce the formation of chaotic phases, the breakdown of stability during the course of any societal evolution.

21. THE FORMAL ASPECT OF RECONSTRUCTION

Utility models under uncertainty and risk have successfully used the paradigm of lotteries to measure utilities and to represent the psychological role of the agents when they solve risky conflicts. In the models, the agents' probabilistic evaluations, their risk attitudes, the influence of the environment, and of random events are expressed as games of chance or lotteries. Mathematically, lotteries are cumulative convex probability functions or combinations of values (prices) with probabilities. Agents may be confronted not only with simple lotteries but, especially in evolutionary conflicts, with compound lotteries or lotteries of lotteries. Formally, these are convex combinations of simple constituent lotteries in the value space. A simple combination of values of alternatives a_i with probabilities is: x_i a_i, where x_i, i = 1, 2, ..., n (Weiss 1988: 115-126). If we combine in our mind the psychological preferential values with probabilities and are willing to risk or avoid a chance, this will enable us to find intuitively in our imagination an optimal prima facie or preliminary solution of future and risky social, political, and economic conflicts. These kinds of lotteries or prospects are exactly our evolutionary ones. They are our inner, mental aspects, our anticipations of our next future. Likewise, prima facie solutions are intelligent, instinctive presentiments of the future course of the societal evolution, which the evolutionary models reconstruct. Prima facie solutions are the final result of our inner optic of an evolutionary path before we realize them. We learn from the lottery paradigm and from its evolutionary stochastic reconstructions that we are not totally guided by blind random events or by the abstract Darwinian survival of the species. On the contrary, when we solve evolutionary conflicts, we are directed by our imagined expected evaluation which we modify simultaneously by our momentary risk attitude. This explains the amazing difference between deterministic economic prognoses, where we follow some lawlike regularities, and expectations, where we are guided by expected evaluations, default rules, and risk.

Prognoses rest on the belief that regularities, like natural laws, will continue into the future, and that random events play a minor role. Prospects, on the other hand, serve as rule-like instructions of how we may optimally solve future conflicts under uncertainty and risk. "Optimal" means here: according to our best knowledge, in unison with our conscience, with responsibility, and with regard to all rules; it also means that we always take into account possible future chances. If we have to evaluate conflicting future evolutionary trajectories, such as future plans, strategies, projects, then it is the best to regard them as special lotteries. To this kind of conflict belong not only the biological conflicts but most of the societal, economic, political, cultural, and, last but not least, according to Wuketits, ethical conflicts (Leinfellner 1993: 38ff; Wuketits 1993: 208ff).

22. SUMMARY

22.1. The psychological aspect

Individuals are not only able to evaluate uncertain chances; with the help of their evaluative abilities, they are also able to put up utility "models" with utility functions for any combination of possible gains or losses. We have enough information about how many gains and losses we may expect in a future lottery. Therefore, we may compare and evaluate quantitatively future lotteries (Allais 1979: 43). Before any realization, we know - like in an ordinary, commercial lottery - how big or how small the possible individual gains or losses will be; we know the value of the whole lottery and our risks connected with the lots to be drawn. It is a further great advantage that we can choose between future prospects (evolutionary trajectories) with the help of the expected utility and our risk attitude. The free combining of possible values with probabilities extends our inner prevision of the future; and our risk attitude selects from the expected solutions the optimal one which stabilizes and secures our individual and societal future. The empirical realization (as canalization or initiation) of strategies or evolutionary trajectories confirms or disconfirms to a certain degree our intended, prima facie solution. If the realization disconfirms the intended prima facie solution and deviates from it, we have to learn how to improve the next solution. For example, we may have miscalculated the rate of accidents of an atomic plant when we compare its utility with the utility of fossil energy.

22.2. The formal aspect of utility

The lottery paradigm instigated the invention of stochastic methods for computing the solution of evolutionary conflicts (Schuster & Sigmund 1985: 255). It led to new theoretical explanations of evolutionary trajectories by the evolution equation, by genetic algorithms, and by dynamic matrices. Finally, it allowed to explain the practical mechanism of how to start (canalize) evolutionary paths in reality. Because of their dependency on our evaluations, on future random events, and on our risk attitude, utility models under uncertainty and risk show clearly why the traditional prognoses so often fail. Traditional prognoses are in risky cases unable to take into account the external role of random events and the internal, psychologically rooted evaluations under uncertainty and risk.

22.3. Evolutionary canalization and the influence model

The lottery paradigm and its theorization in evolutionary game theory explain in a new way the partial influence which we exert on evolutionary processes, for example, on societal evolutionary conflict solutions; it also explains the limits of forecasting. Democracies with their abundance of accepted rules offer us the greatest leeway for free choices among possible evolutionary trajectories. We can choose freely between a variety of admitted projects, societal developments, evolutionary paths, trends, and strategies, all of which lead into the future. For example, we may vote not only for market or ecologically oriented economies; we may also utilize heuristically random events for the introduction of creative (totally new) or innovative (partially new) solutions of societal conflicts. We may even invent or create new societal forms, such as eco-social markets, institutions, and rules. But as regards evolution, the most dangerous sin is to do nothing. Because of the hierarchical stratification of the societal and the underlying biological evolution, we always have to bear in mind that biological solutions may gain the upper hand if we don't solve, or are unable to solve, our societal evolutionary conflicts. The "tribal warfare" in Africa, the racist-terrorist nationalism in European countries, religious fundamentalism of any kind, and terrorist violence in general are warning examples. Therefore, we should not hesitate to delineate evolutionary trajectories, invent new ones, and try to initiate them before it is too late.

22.4. Responsibility

Realization entails accountability and responsibility. Responsibility begins for all of us when we decide for L^1 = 100% A and 0% B in a complex lottery containing L^1 and L^2 and neglect the second possible alternative L^2 = 99.9999% A and 0.0001% B. B is a huge loss and A a comparatively large-sized gain. Or, how much risk, how much harm or losses, for example, unemployment, etc. can be socially tolerated? Finally, we also have the responsibility for a socially just and fair distribution of the gains of lotteries, such as projects, programs, among all persons, not just among the rich. Here more "direct democracy", like in Switzerland and the USA, should be used. After the public has been informed about the risks of realizing solutions and accepted a solution by majority vote, the realization of such conflict solutions should start without delay. Since there are no guarantees of a 100% success, only vigilant constant interventions can help us to secure our societal future in case something goes wrong or when unforeseeable random events occur.

22.5. *Cultural (societal) and biological evolution*

We all know that today our scientific and technological possibilities to interfere in the biological evolution are, firstly, limited. Secondly, in many cases we don't know the great hidden risks of such interferences. This is mainly due to our incomplete knowledge of randomness, of the human and animal genome, and of the evolutionary process in general. The American and the French genome projects are certainly a first step in deciphering the instructions encoded and "written down" in the human genes. The complete knowledge of our genes will reveal the blueprints of our bodies on the one hand; all past experiences, all the instructions or rules of how we have solved our past individual and social conflicts on the other. This tallies with the fact that 90% of all genetic instructions are "behavioral rules", residual "historical" instructions of how we have solved - successfully and unsuccessfully - societal conflicts, roughly until the evolution of language.

We know that our attempts to map and understand all the genetically encoded instructions which are stored in our genes need profound and costly biochemical research, including elaborate apparatuses. But the understanding of that part of our past which has been recorded linguistically, that is, the understanding of written or orally transmitted societal rules of how to solve societal conflicts, requires only less costly ethnographic, historical, and linguistic knowledge. If genes are encoded instructions and rules and the basic units of our biological heritage, then, according to Lumsden and Wilson, culturgens are the corresponding linguistically transmitted instructions and rules (Lumsden & Wilson 1981: 368). Ever since human records began, they have been stored and encoded in our individual and collective oral and written memory, our tradition. They show up "phenotypically" in relatively homogeneous sets of artifacts, behaviors, linguistic mentifacts - and decision rules.

Seen from this new evolutionary sociology, culture is the pool or the sum of all artifacts, forms of behavior, instructions, and mental concepts (mentifacts), transmitted orally and in writing by learning and tradition among the members of a society. Each society's culture is characterized at a certain time by some specific prevalent traits that appear in the majority of its members with a higher frequency than other traits. In more statistical terms, the percentage of the preferred customs, etc. at a certain time will be higher than the percentage of the less preferred, the more idiosyncratic ones. The distribution of the societal properties at a certain time takes the typical form of a lottery: $u(L) = x_1a_1 + ... + x_ia_i$.

Cultural evolution, then, is any change in the culturally transmitted artifacts, forms of behavior, institutions, or mentifacts within and across generations. Statistically, it is the change from one distribution L^1 at time t_1 to another, L^2, at time t_2.

We will not go into the statistical details of Lumsden and Wilson's coevolution theory, a theory which discusses biological and cultural evolution in tandem (cf. also Leinfellner 1991: 229-250). Just one example, to give you some idea: the concept of coevolution entails the fascinating hypothesis that during long periods of cultural evolution favorable and genetically successful societal forms are "written down" into the genome. The genetically inherited memory and the individually and collectively transmitted memory together function like double bookkeeping.

Here we are only interested in the new role of history, of our past individual and collective cultural and societal experiences, and of memory in solving evolutionary conflicts. The economist Marshak was one of the first to call prospects, the simple and compound lotteries, "histories". Thus, he connected their historical, retrospective role with their prospective one. The first aspect enables us to learn from history; the second, to look into the future. In spite of the pessimistic adage that most people "learn from history only that we have not learned anything", good players, decision makers, managers learn eagerly from their past solutions for solving future conflicts. For Marshak (with whom I spent a research semester in Heidelberg), the individually and collectively stored successful rules and methods belong to retrospects. One of the greatest advantages of democratic societies is that they store and transmit historical materials, methods, and rules from generation to generation, while dictatorships tend to destroy or at least falsify them.

23. THE USE OF FORMAL MODELS

The present model has the purpose to demonstrate the dualistic view of the general theory of societal evolution, its phenotypical dynamic, and its psychological aspects. On the one hand, it uses a simplified form of the Eigen-Schuster differential equation of biological evolution (equation (1) below) for the explanation of the external, temporal dynamic of societal evolution. On the other hand, the matrix method represents stochastically the serial conflicts of the same evolutionary track as they change under the influence of psychological preferences, evaluations, lotteries, the expected maximation, and risk.

Within the dualistic theoretical framework, the dynamic differential equations mirror the phenotypical aspect of evolutionary conflict solutions by the temporal increase and decrease and the relative stability of the frequency of a certain societal form (trait, custom, etc.) in a given population. The value matrices formulate mathematically the internal role of the psychological preferences, evaluations, the role of lotteries, their expected values (utilities), the impact of random events, and the psychological-genetic role of risk attitudes when we solve societal evolutionary conflicts. Finally, it will be shown that the two solutions of evolutionary conflicts yield identical solutions which can be explained by the concept of relative stability and by the psychologically and genetically rooted, intelligent abilities of evaluations, the maximation of expected values, and their modification by our risk regulation.

We have compared the matrix method to taking an "evolutionary" lottery ticket again and again if one is exposed to hazard or relatively small random events. In this case, one cannot any longer rely exclusively on regularities or natural laws. Here the function of evolutionary lotteries helps us to understand solutions of evolutionary conflicts in which the likelihood of success or loss depends on internal preferences; on risk and chance, on hope and fear, according to Hagen; on fate; and on our or the others' uncertain choices and their free will in future periods of our lives.

Since the evolutionary gambling situation is always risky, we prefer to go through the possible future alternative trajectories, moves, and possible chances in our imagination before we realize our intended solution of a conflict. Therefore, to internalize a conflict means to anticipate this conflict and its possible solution. Equivalently, learning to solve a serial conflict means to imagine the conflict in our mind, to use our memory of past experiences, to learn the rules, to invent, if necessary, new rules, and to evaluate all possible solutions carefully before we realize them.

Most models of evolutionary lotteries work with stochastic reconstructions of our internal solutions of evolutionary conflicts vis à vis random events. The psychological basis of "internalization" is to imagine all possible future alternative trajectories and solutions by taking into consideration the past and the future possible random events and the future risks of winning or losing. Additionally, most societal conflicts depend on the others' moves and acts. For that reason, we put ourselves into the role of the others in order to find out how they evaluate us and our moves. This is exactly what conflict-solving models express. The matrix method reconstructs the "internalizations" of evolutionary conflicts. Finally, our risk attitude always evaluates the extent to which the others can harm us or be useful for us, etc. Socialization means learning from imagined social and environmental losses or winnings; cooperation begins with putting ourselves into the other's shoes, as the (American) Indian proverb goes. Therefore, cooperation weighs far more than competition (Axelrod 1984: 145; Leinfellner 1991: 244; 1993: 55). Matrix models reconstruct theoretically what is going on in our mind when we solve evolutionary conflicts. They describe the combining of values with possible probabilities, the maximizing of expected utilities and, finally, their modification by calculating automatically the risk involved in each possible and imagined solution. Evolutionary models, then, describe what is going on in our "intelligence" when we face our uncertain future, when we ventilate in our imagination our uncertain solutions of future evolutionary conflicts. These models lay bare the internal, inductive procedure or combining mechanism of solving conflicts under uncertainty and risk ahead of time before we actually realize our solutions. Thus, the psychological foundation of solving conflicts vis à vis randomness in these models rests on our psychological, genetically rooted risk attitude and reveals the scope and limits of our Promethean, mental ability to look into our future.

Incredibly enough, this has all been reconstructed by the matrix method and its models; it may serve as an explanation of the psychological and genetic roots of how we optimally solve evolutionary conflicts. There are no doubts that intelligent players would always prefer to play a game in their imagination or to study an exact model of how to win before they make an actual, practical move. The social impact of the availability of more and more conflict-solving models which we can use before we realize a solution is obvious.

24. APPENDIX: A FORMAL DUALISTIC MODEL OF SOCIETAL EVOLUTION

As we have already seen, the mathematical matrix models reconstruct evolutionary conflicts by taking into account especially all the discussed internal psychological factors and possibilities in order to compute possible optimal solutions. Evolutionary matrix models imitate, therefore, our intelligent, combinatorial, and inductive reasoning when we solve evolutionary conflicts under uncertainty and risk. They represent all the pros and cons in mathematical form, match all the possible combinations of possible values with possible probabilities and risks, and use societal decision rules to compute the optimal solution. This kind of stochastic modelling is a "synthetic-a priori" method: with the help of former experience, the matrix method reconstructs what is going on beforehand in our mind and offers us possible, prima facie, optimal solutions. Likewise, within this method, an evolutionary lottery is a typical game of evaluation under risk "against" or "with" randomness. "Against" means losing competitively against randomness; "with" means that we win "cooperatively" with the help of random events.

The mathematical reconstructions in stochastic models define a lottery $u(L)$ by the uncertain alternatives $u(x_i a_i) = x_1 a_1, x_2 a_2, \ldots, x_n a_n$, $i = 1, \ldots, n$ and $\sum x_i = 1$. The probabilities x_i in the matrix version are in most cases objective ones because the subjective probabilities now show up in the risk factors which modify the expected utilities. Therefore, objective probabilities are frequencies of societal forms in the equational model where $\sum x_i = 1$ and $i = 1, \ldots, n$. In the matrix version, the a_i's represent our values of actions, moves, strategies, of evolutionary "lottery tickets". The probabilities and frequencies depend partially on random events, partially on our acts and the acts of the other participants. The total value of a lottery u(L) is its expected utility, the sum of all alternatives $x_i a_{i.}$. The modification of the expected utility u to the risk utility U yields the optimal solution U(u(L)) of the evolutionary conflict. In compound lottery models, we normally choose one lottery $L_i = x_1 a_1, \ldots, x_n a_n$ from a compound lottery $L^* = y_1 L^i, \ldots, y_k L^k$.

The matrix represents always a cross section of each participant's expectations in an evolutionary conflict at a certain time t. The temporal sequence of solutions resembles a serial dynamic n-person game (Leinfellner 1984: 246-277). The expected utility of an evolutionary lottery u(L) for one player i, promoting a societal form X^i is: $\sum x_i a_i$. The average expected utility for all is $u(L^*) = \sum\sum x_i a_{ij} y_j$. It is clear that the x_i, y_j's change not only by our successful or unsuccessful solutions, by our decisionmaking or realizations of evolutionary serial conflicts but also by the impact of random events.

The evolution equations in this model of a societal evolution represent the most important dynamic aspect of societal evolution, its change in time, more precisely, the differential and temporal changes in the course of an evolutionary trajectory (of a series of conflict solutions), as expressed by formula (2). The changing (increasing, decreasing) or unchanged probabilities x_i, y_j are now frequencies of societal forms, customs, arti- and mentifacts, for example, of a market form X^i = a moderate capitalistic market form, among other forms X^j = ecological market, X^k = socio-ecological market, X^n = regulated market form in a democratic society. The changes dx/dt may be caused by our interferences, such as cooperation, competition, our realizations, and by random events. The matrix represents, therefore, the momentary evolutionary state of societal forms and characterizes at a certain time point t the state of evolution of the whole society. The horizontal columns and the vertical rows are evolutionary lotteries for the participants; they represent their psychic attitude, views, evaluations, and expectations of the future course of the trajectories of their societal evolution.

The general form of a value matrix at the time point t_1 is:

$$
(1) \qquad
\begin{array}{cc|ccccc|}
 & & Y^1 & & Y^j & & Y^n \\
 & & y_1 & & y_j & & y_n \\
X^1 & x_1 & a_{11} & \ldots & a_{1j} & \ldots & a_{1n} \\
X^2 & x_2 & a_{21} & \ldots & a_{2j} & \ldots & a_{2n} \\
 & & \cdot & & \cdot & & \\
 & & \cdot & & \cdot & & \\
X^i & xi & a_{i1} & \ldots & a_{ij} & \ldots & a_{in} \\
 & & \cdot & & \cdot & & \\
 & & \cdot & & \cdot & & \\
X^m & x_m & a_{m1} & \ldots & a_{mj} & \ldots & a_{mn}
\end{array}
$$

The formal model ventilates, on the one hand, the question if and where the solution of any societal evolutionary conflict between different societal forms depends on the conscious realizations (decisions) and interferences of the decision makers, managers, and participants in the society's evolution. On the other hand, it shows when and how societal evolution depends on interferences of random events. Albeit their weights can only be estimated from case to case, this model shows clearly that our influence on our society's evolution is amazingly high.

The attractor of any solution is, as already discussed, the establishment of an evolutionary stability for the individuals and their societies. Mathematically and empirically, any democratic societal form which lasts longer than any other and can "defend" its stability against any other form within its population from generation to generation as well as against small random events is evolutionarily stable. Societal selection now shows up in our model as an increase, decrease, or unchanged state of the frequency x_i of societal forms promoted by their users. One could say that, during the course of the societal evolution, the frequencies of the societal forms indicate their evolutionary rank. Therefore, the temporal change dx/dt represents the evolutionary trajectories, the ups and downs of a specific societal form within one's society. In other words, the increase of the frequency of a societal form will indicate its growth, until eventually it stays stable, and its optimal state is reached. To reach an evolutionarily stable state terminates optimally an evolutionary conflict between the evolving societal forms. The optimal solution will stay stable, provided no further dangerous disturbances occur (see formulation 2).

In this model, any possible interference by decision making (realization of solutions) in the course of the societal evolution will be marked by an arrow, "$\rightarrow$". We should not forget that in the matrix version each cross section of the evolution consists of the expected evaluations of all societal forms a_i or of evolutionary lotteries. The evolutionary lotteries symbolize the possibilities, the chances (winnings and losses) of the realizers and promoters of their own societal form within their population and, at the same time, their psychological evaluation of the ongoing societal evolution.

Finally, the model shows very clearly the manifold ways of human interferences and of random events during societal evolution within democratic western societies.

The temporal flow of the societal evolution and its multiple dependency on evolutionary criteria can be expressed in the n-person evolutionary game dynamic by the following simplified differential equation of Eigen-Schuster type, i.e., (2):

$$(2) \quad dx_i /dt = (E^i - R^i)\, x_i + \Sigma\, \sigma^{ik}\, x_i$$

The following empirical interpretation of (2) lists all possible mutual interdependencies, factors, possible interferences (→) which may have a special and a common impact on the course of a society's evolution which are known today. Interference may be used here for good, bad, and optimal outcomes of evolutionary solutions or decisions by decision makers, managers, and agents in general.

x_i is now the frequency (concentration) of a societal form X^i among other forms. Its expected value is expressed in the matrix version by an evolutionary lottery where each societal form has a certain expected value. For example, the first row in the matrix above expresses the expected utility of a lottery L^1 for a certain user of a societal form X^1. The lottery shows at the same time his dependency on the other participants and on random events, etc.:

$$(3) \quad u(L^1) = y_1\, x_1\, a_{11}, \ldots, y_n\, x_1\, a_{1n}$$

Now, in the evolution equation (2), the factor E^i stands for any increase of reproduction of a societal form i by transmission, that is, conscious or unconscious tradition which interferes in the course of the societal evolution (→). For example, to accept any past optimal solution of an evolutionary societal conflict increases the frequency x_i of a societal form i.

R^i is the second factor in (2). It indicates the diminishing frequencies of societal forms, for example, by political interference, restricting laws, influence of the media, fashions, etc. Both factors, R^i and E^i, are dependent on evolutionary subfactors.

E^i is dependent on the subfactors A^i and Q^i in following way:

$$(4) \quad \rightarrow E^i = A^i \,.\, Q^i \text{ at time } t_{t+1}$$

E^i symbolizes the actual rate of transmission or of tradition of a societal form X^i from generation to generation at the initiating period.

→ Q^i expresses the factor of cooperation (quality factor) which can be increased by education, teamwork, scientific knowledge, and enlightenment. If $Q^i = 1$, no deviations or errors in the transmission will occur.

→ $A^i(1-Q^i)$ is the rate of change of a societal form caused mainly by disturbing random events and errors Q^i. Similar to mutations in the biological evolution, the total number of deviations and of new societal forms deviating from the established one is: $A^i(1-Q^i)$. Here again, the rate of deviations can be reduced (→) by scientific and practical training, technological or preventive measures, and finally by financial support. This means it can be increased or decreased at will.

$$(5) \quad R^i = D^i + K^i$$

This equation shows the dependency of R^i on the subfactors D^i and K^i. R^i is a possible factor of intentional (→) corrections or unintentional deviations (errors) in the transmission of a societal form. D^i is the natural or even biological rate (→) of decomposition or vanishing of a societal form, caused by relatively big random events.

K^i is a naturally or artificially controllable, conscious correction factor. For example, missing competition or cooperation (team work), or a stop of public or financial support for research, etc. may diminish or wipe out societal forms artificially.

In the formula $\Sigma\ \sigma^{ik}\ x_i$, σ^{ik} expresses the transitional probability of the backformation of a societal form X^i from another, very often similar, societal form X^k.

Finally, we get the evolutionary dynamic which depends on the selective interrelationship of the factors $A^i\,Q^i$ and D^i:

If $A^i\ Q^i > D^i$, then the average excess of the natural and/or artificial spreading (growth) of a societal form which is partly caused (initiated) by the optimal realization (decision making or channeling) of an evolutionary solution shows the success of the realization of an optimal solution. One reason can be σ^{ik}, which indicates the probability of an accidental backformation of a societal form X^i from a similar societal form X^k.

If $A^i\ Q^i < D^i$, then this clearly shows the failure of any interference in the evolutionary societal process, of channeling of a specific societal form.

$$(6) \quad W^i = A^i\,Q^i - D^i\,W^i$$

is a typically selective factor; it is based on the excess of growth A^i - D^i of a certain societal form X^i , i.e., the spread of an individual societal form X^i.

From the value-theoretical interpretation in the matrix form, we know that W^i = (L^i); therefore, W^i can be expressed mathematically as the expected utility of an evolutionary lottery, for example by $\Sigma x_i\ a_{ij}$. Now $W^{i}*$ represents the average excess of growth or spread, measured by the expected utility of all lotteries $(L) = \Sigma\Sigma x_i\ a_{ij}\ y_j$ (Leinfellner 1984: 257).

The general default/decision rule for an optimal solution of a conflict, if aversion of risk exists, is given by: if advantageous distributions of random events prevail and/or advantageous societal interference in the societal evolution exist, then:

$$(7) \quad W^i \geq W^{i}*$$

indicates that the societal form X^i has gained the upper hand and stays relatively stable. This evolutionary stability is equal to Maynard Smith's formulation of an optimal evolutionary strategy, (8), which belongs to the matrix method of the stochastic theory of evolution. An evolutionary stable strategy belongs to the optimal solution of an evolutionary conflict if either:

$$(8) \quad x\, a_{ij} e < e a_{ij}\, e, \text{ or, if } x\, a_{ij}\, e = e a_{ij} e, \text{ then } x a_{ij} x < e a_{ij} x$$

holds, according to Maynard Smith (Maynard Smith 1981: 14ff, 198; Leinfellner 1984: 260ff). That means a haphazard societal form arising in small numbers will be less fit than e because it will lose out in any case either against e or against itself. In a perfect democratic society, consisting entirely of individuals adopting the evolutionarily stable strategy e, any emerging rare variants of undemocratic societal forms which adopt a different, undemocratic strategy x would never have a chance to increase in frequency. Hence, a democratic society can stay invariant even in the case of small disturbances by unfortunate random events and normal selective pressure (for more details see Maynard Smith 1982: 56, 204; Leinfellner 1984: 261).

If (6) does not happen, and if random events occur in great numbers, and/or detrimental societal and environmental interference sets in, the societal form X^i may vanish from the society; chaotic phases may begin, as indicated by the following condition:

$$(9) \quad W^i < W^{i*}$$

This indicates that the societal form X^i becomes instable and will cease to exist or go extinct.

This model should give some scientific insights into our abilities to interfere to our advantage, and at the same time, it reveals our limits to influence our societal evolution. Contrary to the pessimistic beliefs of some evolutionists that evolution is the exception and extinction the rule, and that mankind may go extinct sooner or later, just as 90% of all species on earth, this article shows that we already possess the scientific abilities, and soon will possess the technical power, to take our evolutionary fate more and more into our hands. Along with it, we must face the great risk and responsibility as well as the great potential posed by this unprecedented scientific and technical power. Therefore, under the condition that we maintain our democratic welfare state, support our technological advances, and continue to increase our scientific knowledge and research of the evolutionary process, we have the best chance that the optimal continuation of our societal evolution is not an exception and extinction does not become the rule.

Prof. Dr. Werner Leinfellner
University of Nebraska, Lincoln, USA
University of Graz, Technical University of Vienna, Austria
Correspondence: Quaringasse 22, 1100 Wien, Austria

REFERENCES

Allais, M. [1953], "Le comportement de l'homme rational devant le risque: critique des postulats de l'école Americaine", *Econometrica 21*, pp. 503-546.

Allais, M. & Hagen, O. [1979], *Expected Utility Hypotheses and the Allais Paradox* (= Theory and Decision Library 21), Boston.

Allais, M. [1988], "The Machina Theory: A Critical Analysis", in: Munier [1988], pp. 347-403, Theory and Decision Library, Boston.

Axelrod, R. [1984], *The Evolution of Cooperation*, New York.

Bamberg, G., Coenenberg, A.G. [1991], *Betriebswirtschaftliche Entscheidungslehre*, München.

Damasio, A.R. [1994], *Descartes'Error, Emotion, Reason and the Human Brain*, New York.

Dennett D.C. [1991], *Consciousness explained*, Boston.

Eigen, M., Schuster, P. [1979], *The Hypercycle A Principle of Natural Self-Organization*, Berlin.

Friedman, J.W. [1986], *Game Theory with Applications to Economics*, New York.

Gandolfo [1980], *Economic Dynamics, Methods and Models* (2. Auflage), Amsterdam.

Genz, H. [1990], *Symmetrie, Bauplan der Natur*, Piper, München.

Gheradie, S. [1993], "Misère de la prévision", in: *Le Monde*, 19. Okt. 1993, pp. 30.

Götschl, J. [1988], "Wissenschaftlicher Fortschritt und Bedingungen für Humanitätsgewinn", in: *Wissenschaftsforschung Graz*, pp. 9-25.

Grunert, K., Ölander, F. (eds.) [1989], *Understanding Economic Behavior*, Theory and Decision Library, Boston.

Hagen, O. [1983], "Paradoxes and their Solutions", in: *Foundations of Utility and Risk Theory with Applications*, in: Stignum, B., Wnstop, F. (eds.), Theory and Decision Library, Boston.

Hagen, O. [1979], "Towards a Positive Theory of Preferences under Risk", in: Allais & Hagen [1979], Theory and Decision Library, pp. 271-302.

Hagen,O. [1994], "The World of Fear and Hope", in: *Revolutionary Changes in Understanding Man and Society*, Götschl, J. (ed.) in Print.

Holland, J.H. et al. [1986], *Induction*, Cambridge.

Hursh, St. et al. [1989], "The Quantitative Analysis of Economic Behavior with Laboratory Animals ", in: *Understanding Economic Behavior*, Grunert, K., Ölander, (eds.), Boston, pp. 393-407.

Kahneman, D. and Tversky, A., "A Prospect Theory: An Analysis of Decision under Risk", in: *Econometrica 47*, pp. 263-291.

Kroeber-Riel, W. [1975], *Konsumentenverhalten*, München.

Leinfellner, E. [1992], *Semantische Netze und Textzusammenhang*, Frankfurt.

Leinfellner, E. [1994], "The Broader Perspective of Negation", in: *Journal of Literary Semantics*, XXIII/2.

Leinfellner, W. [1981], "Kausalität in den Sozialwissenschaften", in: Posch, G. (ed.), *Kausalität, Neue Texte*, Reclam, pp. 221-260.

Leinfellner, W. [1984], "Evolutionary Causality, Theory of Games, and Evolution of Intelligence", in: *Concepts and Approaches in Evolutionary Epistemology*, Wuketits, F. (ed.), Theory and Decision Library, Boston, pp. 233-277.

Leinfellner, W. [1986], "The Prisoner's Dilemma and its Evolutionary Iteration", in: Diekmann, A., Mitter, P. (eds.), *The Paradoxical Effects of Social Behavior*, Heidelberg.

Leinfellner, W. [1987], "Die Entscheidungstheorie - eine Konfliktlösungstheorie", in: *Zeitschrift für Wissenschaftsforschung 4*, pp. 15-31.

Leinfellner, W. [1987], "Evolutionäre Erkenntnistheorie und Spieltheorie", in: *Die Evolutionäre Erkenntnistheorie*, Wuketits, W., Riedl, M. (eds.), Berlin.

Leinfellner,W. [1983], "Evolution of Intelligence", in: *Epistemology and Philosophy of Science*, Weingartner, P., Czermak, J. (eds.), Vienna, pp. 161-167.
Leinfellner, W. [1984], "Evolution und Schöpfung:Ursprung des Universums", in: *Philosopie der Religion*, Gombocz, W. (ed.), Wien, pp. 135-147.
Leinfellner, W. [1986], "The Prisoner's Dilemma and its Evolutionary Iteration", in: *Paradoxical Effects of Social Behavior*, A. Diekmann & P. Mitter (eds.), Heidelberg, pp. 135-148.
Leinfellner, W. [1988], "Nature is the Real Designer", in: *Austria Today 2/88*, pp. 39-43.
Leinfellner, W. [1989a.], "The Change of the Concept of Reduction in Biology and in the Social Sciences", in: *Centripetal Forces in the Sciences*, Radnitzky, G. (eds.), New York, pp. 55-77.
Leinfellner, W. [1989b.], "La Théorie des Probabilités et L'analyse des Séries Temporelles", in: *Ann. des Mines*, pp. 33-39.
Leinfellner, W. [1991], "Traditional Ethics, Ethical Decision Theory and Evolutionary Ethics", in: *Absolute Values and the Reassessment of the Contemporary World*, New York, pp. 229-250.
Leinfellner, W. [1993], "Ein Plädoyer für die Sozialethik", in: Lütterfelds, W. (ed.), *Evolutionäre Ethik zwischen Naturalismus und Idealismus*, Darmstadt, pp. 32-65.
Lütterfelds, W. (ed.), *Evolutionäre Ethik zwischen Naturalismus und Idealismus*, Darmstadt.
Lopes, L.L. [1988], "Economics as Psychology: A Cognitive Assay of the French and American School of Risk Theory", in: Munier, B. 1988, Theory and Decision Library, Boston, pp. 405-415.
Lumsden, Ch., Wilson, E.O. [1981], *Genes, Mind and Culture*, Cambridge.
Munier, B. [1988], *Risk, Decision and Rationality*, (= Theory and Decision Library), Boston.
Maynard-Maynard-Smith, J. [1982], *Evolution and the Theory of Games*, Oxford.
McConell, [1966], *Economics*, New York, Russian edition 1993.
Mesterton-Gibbons, M. [1992], *An Introduction to Game-theoretical Modelling*, New York.
Morgenstern, O., Neumann, J. von [1961], *Spieltheorie und wirtschaftliches Verhalten*, Würzburg.
Nida-Rümelin, J. [1987], *Entscheidungstheorie und Ethik*, München.
Pope, R.E. [1991], "Lowered Welfare Under the Expected Utility Procedure", in: Chican A. (ed.), *Progress in Decision, Utility and Risk Theory*, Boston, Theory and Decision Library, pp. 125-133.
Prigogine, I. [1982], *Vom Sein zum Werden*, Hamburg.
Rapoport, A. [1980], *Mathematische Methoden in den Sozialwissenschaften*, Würzburg.
Rumelhard D.E. and McClelland, J.L. [1986], *Parallel Distributed Processing, Exploration in the Microstructure of Cognition*, Cambridge.
Schuster, P., Sigmund, K. [1985], "Towards a Dynamics of Social Behavior: Strategic and Genetic Models or the Evolution of Animal Conflicts", *J. Social Biol. Struct. 8*, pp. 255-277.
Sen, A. [1970], *Collective Choice and Social Welfare*, London.
Selten, R. [1988], *Models of Strategic Rationality*, Dordrecht.
Simon, H.A. [1956], *Theories of Decison Making in Economics and Behavioral Sciences*, American Economic Review.
Tulving, E. [1983], *Elements of Episodic Memory*, New York.
Weiss, M.D. [1988], "EU Without Continuous Preferences", in: Munier 1988, pp.115-126.
Wuketits, F. (ed.) [1984], *Concepts and Approaches in Evolutionary Epistemology*, Theory and Decision Library, Boston.

RISK IN UTILITY THEORY, IN BUSINESS AND IN THE WORLD OF FEAR AND HOPE

OLE HAGEN✝

ABSTRACT

Practically all decisions are made under uncertainty about the consequences of the alternative actions we can take. Sometimes we don't even know the probability distributions over the possible consequences, given the action. The history of basic decision theory is outlined and it is shown how it develops through "paradoxes," i.e. observations from "real life" and real or mental experiments that run counter to what should be expected from received theory. It is shown how such paradoxes, including for example the notable "Allais Paradox," invalidate the "expected utility theory" as a descriptive/predictive theory, and also that it is, in its present form, ambiguous and useless as a norm for "rational" decisions.

Allais' theoretical foundation of an alternative approach is summarized. Its intellectual descendants are surveyed and exemplified by a model where the utility of a game is a functional of the expectation, distribution and skewness of the probability distribution over utilities of outcomes. It is shown how it is extended and modified by later research. Allowance is, for example, made for how fear of the regret when the consequence is known can influence a decision.

Uncertainty permeates all fields of human activities, and basic decision theory should be adopted in all social and behavioral sciences.

Keywords: Risk, uncertainty, utility, paradoxes, games, regret.

INTRODUCTION - RISK, UNCERTAINTY, VALUE

In everyday language "uncertainty" is used to characterize the aspect of a situation where we are not certain of what will happen, whether dependent or independent of our actions, whereas "risk" describes the negative side of what may happen, more or less synonymously with danger.

In decision theory, a usage has developed that "risk" is used when the probability distributions over the possible events and consequences of our actions are known. "Uncertainty" means that they are not. This use of common language words as technical terms may be confusing: possessing a lottery ticket is not sensed as risk. In this article the word risk is used in the technical sense in technical contexts, elsewhere as in everyday usage. Uncertainty includes risk, and when probabilities are unknown this is indicated explicitly. Also, uncertainty is to be understood as embracing the

J. Götschl (ed.), Revolutionary Changes in Understanding Man and Society, 191–210.

emotions of tension, fear and hope. Let us, at this stage, consider a practical example. Cathy, an old-fashioned girl, does not invite boys as escorts to a ball but hopes to be invited. To the coming ball she would be pleased to accompany either Adrian or Bill, but would prefer Adrian. Now, before Adrian acts, Bill invites her. She cannot stall Bill, she must accept or decline.

A theoretical analysis of this case is difficult. We don´t know the probability that Adrian will invite her, and we have no numerical values of the two boys. Actually, we have similar difficulties in business, private life, politics and war. No wonder the first decision theorists preferred discussing games where probabilities are known and outcomes are gains and losses in cash.

THE ST. PETERSBURG PARADOX

In the 18th century intelligent and well-educated persons took it for granted that in choosing among monetary games with known probabilities you preferred the game that gave the highest monetary expectation, product sum of probabilities and outcomes.

Before we turn to how this was rejected as a positive theory, let us be aware of its merit as a normative model: it is unambiguous. Given that the possible outcomes and their probabilities are known, there is a unique solution. This may not be a sufficient condition for a good decision norm, but there is a strong case for its being a necessary condition.

D. Bernouilli (1738) constructed a game which, simplified, had the following rules: the participants are A, the gambler, and B, the banker. A fair coin is tossed. If it comes up heads, B pays A \$2. If it comes up tails, it is tossed a second time. If it comes up heads, B pays A \$4 also ad infinitum. The expected prize,

$$E = 1/2 * \$2 + 1/4 * \$4 + \ldots \text{ ad inf.} = \$1 + \$1 + \ldots \text{ ad inf.}$$

According to current theory, the cash value of this game should be infinite. But the fact was that nobody would pay more than a few dollars worth for the game.

The expectation theory was rejected. So the consensus at the time and for more than two centuries. Karl Menger (1934), the mathematician and philosopher (not to be confused with his father, Carl Menger, the economist), pointed out that B could not stand for his obligation. At some point in the geometric series of possible prizes he would be broke. So the expectation of the prize actually paid out was finite. But, Menger pointed out, even a truncated St. Petersburg game, allowing for B's limited resources, could not be sold for its expected value. So the conclusion of

Bernoulli's thinking might still be valid. Bernoulli suggested that games were valued as the expectation of the logarithm of the prizes. This solved for the time being the St. Petersburg paradox. Bernoulli thus foreshadowed the expected utility theory in so far as the valuation of games was the expectation of a function of the monetary prizes. Since the derivative of ln x is 1/x, he also foreshadowed the Austrian concept of diminishing marginal utility and, in fact more directly, Allais' measurement of cardinal utility showing a utility function roughly as the logarithm of total net worth.

THE EXPECTED UTILITY THEORY

The original concept of utility in economic theory was "cardinal", though in a sense somewhat different from what we understand by cardinal numbers. The zero point and the unit could be changed without changing its economically relevant characteristics. It was, in other words, determined up to a positive linear transformation; one could say it was defined on an interval scale, implying that ratios between differences in utility over intervals - even if none is contained in the other - are meaningful. It was mentioned by Pareto (1916) that a positive (not necessary linear) transformation of a (one period) utility function will express the same preferences. This made some economists jump to the conclusion that a merely ordinal preference function would suffice for all purposes. The cardinal utility concept was redundant, maybe meaningless. Unfortunately, they termed their preference function utility, namely ordinal utility. By the time of the second world war ordinalism was dominant among economists.

More or less parallel with this development there emerged a theory of decisions under risk which came to be named expected utility theory. We shall use the presentation of von Neumann and Morgenstern (1953) as reference for this theory. The essence of the theory can be exemplified like this: given that a person ranks three possible situations in order of preference A, B, C, we attach the value 1 to A, and 0 to C. If the person is indifferent in a choice between B and a game with the probability p for A and 1 - p for C, then his utility of B is defined as p. The theory is that in any choice between games whose outcomes are A, B or C the person prefers the one with the highest expected utility. It is implied in the theory and also in one of the "axioms" used to develop the theory, that, if the game:

$$\{Pr.(B) = r, Pr.(C) = (1-r)\}$$

is found equivalent to the game:

$$\{Pr.(A) = qr, Pr.(C) = (r - qr)\}$$

then the utility, u, of B, u(B) is found by solving the equation:

$$r\,u(B) = qr\,u(A) + (r - qr)\,u(C)$$
$$u(B) = q\,u(A) + (1 - q)\,u(C)$$

and given the chosen utility values of A and C:

$$u(B) = qu$$

According to the theory u(B) could have been defined in this way in the first place, and if the theory fits the facts:

$$p = q$$

But will they be equal? We shall return to this later. At this stage let it suffice to point out that the originators of the theory admitted (op. cit.) that their model did not include a "*specific utility of risk*".

At first the N-M theory got a mixed reception. Their utility or "utility index" was in fact cardinal in the sense that a non-linear transformation would distort the preference structure it expressed. They were "accused" of reviving the cardinal concept of utility that was supposed to be dead. Among their critics was Baumol (1951). At a later stage however he found a formula for reconciliation between expected utility theory and ordinalism. The utility index did not express how much one *game* was preferred to another, compared to other preferences, it only indicated the order of preference between *games* (Baumol, 1958). This pretext for a "political marriage" (Hagen, 1988) was very thin. The ordinalists swallowed the fact that, if, of three situations, B is preferred to a 50-50 game between A and C, this meant by definition that the utility difference between B and A was less than that between C and B. But this also implied that, if the games A', B', C' were equivalent to A, B, C respectively, then the corresponding preference could be predicted. Why? Of course because the utility differences in the case of A', B', C' have the same relations as in the case of A, B, C.

The formula was, however, generally accepted, and it became a criterion for being "with it" to be both an ordinalist and an expected utility maximizer. The reservations of the originators concerning the specific utility of risk were soon forgotten. Their "utility index" was interpreted as an expression of attitude to risk. It was forgotten that in the original presentation no special role was attached to money. When the utility index was presented graphically, with money on the horizontal axis and utility on the vertical axis, a person was a risk averter if his curve was concave (seen from below) and a risk seeker if it was convex. Or, if a person could accept a game even if it had a negative expectation, he would be a risk seeker; if he would always prefer the game with the highest monetary expectation, he would be risk neutral; and if he would not accept a game unless it had a positive expectation, he would be a risk averter. In a world without money this distinction between classes of humans, so defined, would be meaningless. If we assume that a person would know what he would buy for a prize, a change in relative prizes could turn a risk seeker into a risk averter and vice versa. But if a cardinal utility was assumed, then a maximizer of the expectation of that utility could be defined as risk neutral, and it could be conceptually possible to define risk seeker and risk averter in analogy with the definitions relative to money.

THE INSURANCE-LOTTERY AND THE LOTTERY-BET PARADOX

The believers in the expected utility principle had a bad moment when it was pointed out that on the one hand nearly everybody bought insurance for their property and most of them also bought lottery tickets. This was paradoxical because an insurance customer pays, in addition to expected compensation, administration, cost and profit for the insurance company; the game of not insuring having a positive expectation, but being risky, is declined, so he is a risk averter. The lottery customer pays, in addition to expected prize cost and surplus for the lottery, so he is a risk seeker.

Friedman and Savage (1948) apparently solved this paradox by constructing a utility curve which was concave to the left and convex to the right, with the inflection point near status quo. This was generally accepted, but the solution was only apparent. Or, put differently, it leads to a new paradox. We must assume that the first prize in lotteries is on such a level that most of the potential customers would rather have it there, than at half the amount with the double probability. If people were maximizing the expectation of a utility index that was convex between status quo and status quo plus first prize, those who win as second prize a substantial amount, but not more than half of the first prize, would be happy to enter this in a 50-50 double or nothing bet. Nobody does (Hagen, 1991). This *lottery-bet paradox* was not pointed out at the time, but a new paradox conquered the scene.

THE ALLAIS PARADOX AND THE THEORY BEHIND IT

Originally presented in 1952 in French and in 1979 in English (Allais 1952, 1979a), it is presented here in terms I find practical.

As a mental experiment consider two choice situations: choice 1 between games A and B and choice 2, games X and Y, as below, games in the form (pzobability, prize; probability, prize):

Choice situation 1:

A: (1, 1M)

B: (0.01,0; 0.89,1M; 0.10,5M)

Choice situation 2:

X: (0.89,0; 0.11,1M)
Y: (0.90,0; 0.10,5M)

M was an amount of money that to most people would be a substantial gain.

This was presented to a number of persons, and many answered that they would prefer A to B and Y to X.

A close look at the four games reveals that moving from A to B as well as from X to Y, means exchanging 11 % probability of winning 1M for 10 % probability of winning 5M, so, if E [u(G)] is the expectation of the utility of outcome of a game, then:

$$E[u(A)] - E[u(B)] = E[u(X)] - E[u(Y)] = 0.11\, u(1M) - 0.10\, u(5M)$$

This means that a person maximizing the expectation of the utility of the outcome, regardless of how the utility curve bends and twists, must either prefer A and X or B and Y, not possibly A and Y, and incidentally not B and X, which rarely occur. The fact that the combination A, Y is frequent and B, X extremely rare, which is well established, is very remarkable. All the more so, since, if a person is faced with a two step game where the first step is a 50-50 chance fork between the choice between A and B and between X and Y, and asked to make his choice in advance, the two combinations have the same expected utility.

The theory behind the above and other "Allaisian" paradoxes, found by Allais and many others, is that the above-mentioned *specific utility of risk* (v. Neumann and Morgenstern, 1953) indeed exists and that it depends on the whole shape of the probability distribution over utilities of the outcome, in particular the standard deviation. Implied is a complementary effect in the probability of a desirable outcome. The effect of a given increase in the probability is the stronger the closer it is to unity (Allais, 1952, 1979a).

For example: moving from B to A changes the probability of winning at least M from 0.99 to 1.00. Moving from Y to X, changes the same probability from 0.10 to 0.11. In both cases the probability of winning 5M in lieu of 1M decreases from 0.10 to zero. It is easy to see that the complementary effect could explain the Allais paradox.

THE POST-ALLAISIAN ERA

After Allais (1952, 1953) there was a period of thundering silence concerning the doubts about EU theory until the publications of Krelle (1968) and Hagen (1969).

Krelle pointed out very clearly that there are two factors at work in the motivation of choices under risk, utility and risk attitude. He expressed this as the choice function under risk being a function of the utility function. This line of thought has been developed further by Bitz (1976). It is a different approach from the one I am partial to, but the presentations are certainly elucidating and touch on the core of the matter.

Springtime came in the seventies and eighties and "a hundred flowers" blossomed in the garden of alternative risk theory. Let me quote Sugden (1989):

For many years almost all economic analysis of choice under uncertainty was based on expected utility theory. The validity of this theory, both as a prescription as to how people ought to choose and as a description of how they do choose, was hardly questioned. The few sceptics, led by Maurice Allais, were generally perceived as eccentrics, outside the main stream of economic thought. Now it seems that their time has come. There is a rapidly-growing literature criticizing expected utility theory and suggesting alternative approaches. Let me say at the outset that this paper does not purport to be a general survey of this literature. [Two excellent surveys are available, written by Schoemaker (1982) up to 1967 as far as Europe is concerned? - 0. H.] and Machina (1983). Rather it describes one class of alternative theories that has been developed in the last few years by, among others, Bell, Chew, Fishbum, Hagen, Loomes, MacCrimmon and myself. I believe that this work provides one promising avenue for progress but it is certainly not the only one.

His list of excellent surveys could be updated with, e.g. Sugden (op. cit.) and other articles in Hey and Lambert (1987) and Boehren (1990).

It would be too much history if I tried to give a fair presentation of all these models. It seems acceptable first that I stick to the "class" defined by Sugden, second that I use my own model as a representative for this class, by the right of seniority and since the difference between models often is more a question of which form of expression satisfies the authors' tastes than of essential differences in content.

HAGEN'S THREE MOMENTS MODEL

The essence of the first version of the model in Hagen (1969) was the same as that of Allais. The definition of the absolute preference was formulated in terms of utility instead of money. It postulated the existence of a cardinal utility u, and in relation to money, x, the function u(x) had the derivatives:

$$u'>0,\ u''<0,\ u'''>0, \ldots$$

An important factor in the evaluation of a game is the expectation of ex post cardinal utility, also the tension caused by dispersion as such. Hagen (1969) puts particular weight on the strong psychological motivation of hope and fear. Therefore, the skewness of the distribution was important. Even a small probability for something very good gives a person hope that can help alleviate the suffering of the dreariness of a grey life. On the other hand, even a small probability for a disaster (compared to the expectation) would cause fear, which means suffering in itself even if the possibility of disaster never materializes.

In Hagen (1969) I did not fall into the trap of distorting the probability distribution itself. Skipping details, the essence of the mathematical formulation was:

$$U(u,s,z)=\max$$

where

U = utility of a game,
u = cardinal utility of outcome,
u = mean of u,
s = standard deviation,
z = a special measure of skewness = m_3/s^2
(except, of course, when $s = 0$ and $m_3 = 0$; then $z = 0$),
m_3 = third central moment

and

$$\delta U/\delta u>0$$
$$\delta U/\delta s<0$$
$$\delta^2 U/\delta s^2<0$$
$$\delta U/\delta z>0$$

The special measure of skewness is chosen because it has an important characteristic: in a two-outcome game it increases in absolute value as the smallest of the probabilities decrease (Hagen, 1979).

When u = u(x), and x = monetary gain/loss then odd number derivatives of u are positive and even number derivatives are negative.

It has been shown that this model absorbs the well-known paradoxes of that time. In fact, it solves all the paradoxes mentioned above with the exception of the juxtaposition paradox.

Werner Leinfellner suggested that my model in Hagen (1969) might be lacking an axiom which captures the effect if outcomes undergo the same change. It seemed reasonable to me to assume that, if they all had the same change in utility, then the value of the game would change with the same amount. I added an axiom stating that, if all outcomes were improved by the same amount of utility, the utility of the game would increase by the same amount. This made it possible to separate the expectation of the ex post utility and the specific utility of risk in an additive function (Hagen, 1972) as follows:

Value of game is $U = E(u)+f(s,z)$
where $f(s,z)$ represents the specific utility of risk.

Obviously, if all u values increase by the same amount, then E(u) increases with that same amount, and f(s,z) remains constant; as a result, U increases with the same amount.

If we make the mental experiment that both cardinal utility exists and that the von Neumann-Morgenstern utility index exists (no specific utility of risk), then f(s,z) evaporates, and the two utility concepts must be identical.

Allais has (1979b) given this axiom the name: *The axiom of cardinal isovariation, and* applied it in a similar way. It is perhaps also worth noting that any *linear* operation on u implies the same linear operation on U.

The 1972 version (Hagen, 1972) contains some theorems that concern games about money which can be tested against reality, in experiments or in observation of market behavior.

It has been pointed out (Sugden, 1987) that some points made by other alternative models can be seen as accommodated within the three moments model.

I quite agree, but must also admit that in other cases this is not so, for example in the case of regret. Technically, regret is the difference (if positive) in the utility of the wealth after the game, had you chosen the action that, after the game, is known to have given the best result, and the result of your actual choice (see for example Loomes and Sugden, 1985). I have therefore adopted some of these recent results in my model in a simple way: the utility symbol "u" is defined as the utility of obtaining the outcome, i.e. the utility of the change in wealth as well as the psychological effect of the experience of the outcome, such as regret, namely the juxtaposition effect.

One might then say that while the function f expresses the anticipation at the time of the decision-making of the emotional experience of tension, hope and fear during the time till the result is known, the variable expresses the anticipation of the emotional experience of learning the outcome.

The effect of time, which has been discussed explicitly by Pope (1985), is a very difficult question, which has been mostly ignored in the literature. One reason may be that people's attitudes differ too much to invite a general treatment in a formal model. One example: there are lotteries of similar structure where in one case the customers wait for weeks to learn the result, and in another the buyer scratches the ticket to learn the result. In one case the gambler can enjoy the feeling of hope (there is a real possibility that he may get rich quick) for a long time, and in the other case he can satisfy his impatience. Both are sold.

JUXTAPOSITION EFFECT

The outcome of a game depends on the state of the world ex post game, for example which number turned out in the drawing of a lottery. But an adequate definition of a state of the world must also contain all other aspects that may be relevant to your choice. If you have no gas mask and no champagne and have to choose between two lotteries where a gas mask and a bottle of champagne are the two sole prizes, you know that the utility of each prize (at least the mask) depends on whether there will be a gas attack or not. We have a case of state dependent utility. Sometimes it depends on what would have been the situation if the decision maker had made another decision, given of course that this is known. The concept of regret, as defined above, is an example of this.

I believe that most readers may have assumed that in each of the two situations of the Allais paradox the two games are stochastically independent.

Let us imagine two roulettes with 100 numbers operated by two croupiers, one for each game, and only the one for the game chosen will be spun. If in situation 1. the decision-maker prefers A, there is no telling what the outcome would have been if he had chosen B. If he prefers B, he will know that, whatever the outcome is, the outcome of the alternative choice would have been 1M, so if he wins nothing, there is regret = u(1M) - u(O). In situation 2, whatever his choice and whatever the outcome, there is no telling what the alternative choice would have been. So there can be no regret. When the games are stochastically independent, the possibility of regret favours A in situation 1 and is non-existent in situation 2.

Let us now imagine that in each situation there is only one roulette which is spun, whatever the choice. We further assume that the distribution of results is fixed prior to the decision-maker´s choice. In situation 1, if A is chosen and the roulette shows that B would have given 5M, then regret = u(5M) - u(1M). If B is chosen and the outcome is 0, the regret is u(1M) - u(0) as in the case above. In situation 2 the possible regret effects depend on how the probabilities are distributed over the 100 numbers. There is an extremely large number of possibilities giving the required probability distributions. We will consider two subsets, each having a large number of elements, the subset defined by a common characteristic.

Subset S: 89 numbers give the outcome zero in both X and Y. Of the remaining 11, one gives 1M in X and zero in Y, and the remaining 10 numbers give 1M in X and 5M in Y.

Subset H: 79 numbers give the outcome zero in both X and Y. Of the remaining 21 numbers 10 give zero in X and 5M in Y. The remaining 11 numbers give 1M in X and zero in Y.

I should think that most readers who might prefer X under S will intuitively prefer Y under H, but let us be more academic:

If we compare situation 1 with situation 2 under subset S, it is clear that, whatever the regret effect might be, it would favour either A and X alike or B and Y alike. The expected utility theory is then not rejected, but do we get a hint as to which of the two expected utility consistent choice combinations is preferable? If we apply the widely accepted rule of minimizing the maximum regret as a supplement to expected utility maximization, then A and X are favoured by the minimax regret rule if [u(1M) - u(0)] [u(5M) - u(1M)], which is at least possible. If however we compare X and Y under subset H, then X is favoured if and only if [u(1M) - u(0)] < [u(5M) - u(0)], which is impossible; so under subset H, minimax regret favours Y.

One of the persons whom Allais asked to choose between the games in his paradox was an outstanding expected utility theorist, L. J. Savage. He made the "paradoxical" decisions, A and Y. When Allais told him that he had violated the expected utility principle, he was reluctant to admit it, but later phoned Allais and said: 'You were right but I was wrong.' He has explained (1979) how he revealed and corrected his "mistake." He did this by arranging the games in such a way that in both choice situations of 11 numbers that in one game gave 1M in the alternative game one gave zero and the other 10 gave 5M. Since in both cases the remaining 89 numbers had the same outcome in both games, according to Savage's sure thing principle, the choice was in fact determined by the preference between the two sets of outcomes over the 11 numbers, so the choices must be either A and X or B and Y. Given that M is a substantial prize, the choices will be A and X.

Again the expected utility principle was assumed to be saved. It was later pointed out that, if instead of the 11 numbers one chose 21 numbers where in the case of the games X and Y the numbers giving a prize were juxtaposed with the numbers giving zero in the other game (Hagen, 1972, 1979), the expected utility principle could be saved (MacCrimmon, 1979; Pope, 1985). The two juxtapositions are elements in subsets S and H above. We were in fact faced with a new paradox foreshadowing the discussion of state dependent utility and juxtaposition effect.

ARE THE SHORTCOMINGS OF THE EXPECTED UTILITY THEORY IMPORTANT?

Some people ask: 'Should a simple, consistent and beautiful theory be rejected just because a few examples show that it does not cover every corner of reality?' Let us see if this question relates to the expected utility theory.

Practically everybody will buy insurance for property they might possess, which can easily be explained as maximizing the expectation of a utility function whose graph is concave (seen from below) or from e.g. the alternative model presented above. The overwhelming majority of these will occasionally buy lottery tickets, which can be explained by a convex utility curve over the relevant area, but will, if put to the test, not enter a 50-50 fair bet on a substantial amount within the same area, which disproves that explanation. But the alternative model explains it. This alone proves that a majority of the population when in a real, practical situation faced with decisions under risk with known exact probabilities follows a pattern systematically different from expected utility maximization.

Let us elucidate the juxtaposition effect through a real life case. The managing director of the Swedish Volvo company some years ago made a tentative deal with the Norwegian government. Volvo would get an oil drilling concession and in return establish a plant in Norway. The deal would have to be confirmed by Volvo's stockholders' meeting. Now the leader and trustee of an investment society did not like the deal and convinced the majority, so the deal was not approved. Then, some time later, the concession reserved for Volvo in the tentative deal turned out to be unexpectedly rich. This was a clear case of regret from the point of view of the members of the investment society, who unseated their chairman. Suppose now that the deal had not specified the concession, but left it open with similar ex ante promises. Obviously the leader of the investment society might still sit in his chair. Had he in advance given more attention to this particular aspect, he might not have dared to oppose the deal, though the expected utility of wealth would be no different in the two situations.

Probably cases like this occur all the time in business. The alternative models are much more in touch with real life than the expected utility theory which should be confined to the ivory tower where it belongs.

NORMATIVE DECISION THEORY

People with a strong emotional attachment to the expected utility principle who reluctantly realize that it does in fact not describe people's, even intelligent and educated people's, actual preferences, have a second line of defence. The theory does not pretend to describe actual behavior but is seen as a norm for rational behavior.

The first necessary condition for a normative theory to be useful is that it is not ambiguous. The rule to maximize monetary expectation meets this requirement. If all monetary outcomes are positive, the same goes for maximizing expectation of logarithms.

Taken literally, expected utility maximizing means: "When choosing between games, maximize the expectation of the utility index which is defined as the one you maximize the expectation of when choosing between games!" This may look like a piece of flippancy but is true. And with a little change it can make sense: let the normative rule apply to *complex* games and the definition refer to *simple* games. For example, two outcome games where any non zero probability is divisible by 1/3 or 1/4. It has been indicated that the proper domain of the descriptive theory should be restricted in this way (Morgenstern, 1979).

Understood this way, using expected utility maximizing could be a useful device if the definition were unambiguous. This question is raised in the presentation of the expected utility theory above. The definition of utility implies that if the certainty of a certain amount of money is equivalent to the probability p of winning a given, higher, amount, given the utility 1, then the utility of the lower amount is defined as p also, if we replace the certainty of the lower amount with the probability r, and p with qr and find the q that gives equivalence, according to the theory p = q. And that is the big question.

Morgenstern's statement (1979) has largely been ignored by his followers. It seems as if in this particular respect I am his best pupil. The statement was made in connection with criticisms of empirical investigation made by Allais and by me. It was directed against too large prizes, too small probabilities and, in my case, that "Hagen´s school teachers" maybe were not so knowledgeable in the theory that they could appreciate the problems. Even though I did not agree, I conducted a new investigation. The respondents were the academic staff of the Faculty of Mathematics and Natural Science at the University of Oslo. The prizes were NOK 10 000 and 24 000 (1 USD approximates 7 NOK), and the probabilities were 1 and 0.5 in one question and 0.5 and 0.25 in the other (r = 0.5). Interpreted in terms of the presentation above, the outcome showed $p > 0.5$ and $q < 0.5$ for the same amount of money. The preference for the "less risky" alternative were 92.3 % when probabilities were 1 and 0.5 and 63.3 % when probabilities were 0.5 and 0.25. The significance test gave 2.969 compared to the 95 % fractile 1.645, see (Hagen, 1979). (To avoid any misinterpretation Morgenstern was given the opportunity to comment in the same volume, but preferred to do so at a later stage. This did not materialize due to his death, nor has anyone else felt called upon to follow up. It is more efficient to ignore this.) There is a large number of (better known) empirical investigations showing inconsistencies (in the definition of the Neumann-Morgenstern utility index or in respondents' perception, as you please). Let me just indicate two that are easily available, in the same book: (Allais, 1979b) and (MacCrimmon and Larsson, 1979).

Some business consultants think they use the expected utility model to help clients make decisions according to their subjective utility. In a sense they may be right in so far as they use one of the infinite number of elements in the set of Neumann-Morgenstern utility indexes. It would always be easy to find another that would contradict them. The theory does not specify which index to use. In the case of the Allais paradox it says the two choices are inconsistent, but it cannot say which is wrong.

It has been maintained that violation of the expected utility principle must lead to reduced probability of financial survival. Now, if the utility function has only two values, survival (1) and ruin (0), this would be right if considering overall strategies. But this would also be a violation of every possible decision rule in the book. However in at least one case (Karni and Schmeidler, 1986) the violations have been characterized by "such as the Allais paradox." It has been shown that in reasonably realistic scenarios under varying circumstances maximizing probability of survival implies the Allais paradox and similar paradoxes (Hagen, 1992).

Having established that expected utility can not provide a unique solution to choice under uncertainty, and being aware that the suggested alternative descriptive theories do not claim nomative validity, one might wonder:

Is it possible at all to establish a complete and unique normative solution?

Allais' *"préference absolue,"* the well-known principle of stochastic dominance of the first order and my own version, where utility takes the place of money, concern the situation where the cumulative probabilities in two games consistently favour one game over the other. This could come close to universal acceptance, but it is not complete. Mostly the curves of cumulative probability intersect, so the preference must be what Allais named *"préference relative"*.

I am inclined to believe that the answer to the question is no.

This does not mean that decision theory cannot contribute to developing a decision technique that can be used by a consultant to help a decision-maker to select among complex games the one that best satisfies his subjective tastes.

One such method is "the proxy method" (Hagen, 1983, 1991, 1992). For each of the actual games the mean variance and third central moment is found, and a "proxy," a two outcome game, is constructed with the same three characteristics. The decision-maker is asked which of these proxies he would prefer. He is then advised to choose the corresponding original game. It has been tested whether there will be a correlation between the ranking of three outcome games and their two outcome proxies with a favourable result (Blasche and Dorfner, 1987). Though positive proof for more complex games is impossible, this is an indication that the method is sound. At least it stood up where it could have been discredited.

THE "REAL" WORLD

Of course the researchers´ activities such as reasoning, model construction, mental experiments and laboratory experiments are part of the real world. But because some of these activities deal with models of other parts of real world activities aimed at understanding and perhaps influencing them, we speak of these other parts as the "real world" as opposed to the model world.

We have so far discussed lotteries which are closely related to our simple models when prizes and probabilities are known. We have also discussed insurance and a special example of a business decision. The best known area of business application of risk and utility theory is portfolio theory. Usually theories in this field are presented as implications of the expected utility theory, which may be true if one forgets the particular shape of the utility function suggested by Friedman and Savage (op. cit). However, a useful model as the one assuming portfolios to be ranked by a function of expectation and standard deviation is a direct mathematical implication of assumptions of reasonably approximate realism (Hagen, 1985).

Let us take a look at business in general.

A merchant, even a small time merchant, cannot run his business without uncertainty. He buys commodities in order to sell them at a profit. He runs the risk that the goods may deteriorate or for other reasons be unsalable at the planned price. If he gives credit to customers, this may cause loss.

A manufacturer has the risk that his machinery may become obsolete and thereby his plant uncompetitive. If he increases his capacity, he may risk that due to unforeseen changes in the market he may not sell enough to cover his increase in fixed costs. On the other hand, if he sticks to his capacity, there is always the possibility, if there are economies of scale, that others through expansion might be able to undercut his minimum price.

In traditional microeconomic theory, as taught in universities, the underlying assumption is more or less that everything is known, not only about conditions when production decisions are made but also when the products are marketed. Optimal production is where marginal cost = marginal revenue. If the price is then lower than the all inclusive unit cost, the plant will by and by be closed down. If the price is also lower than unit cost exclusive of fixed cost, production is stopped. In fact this is not how plants are usually closed. Rather they are closed because of the owner's financial insolvency. It can be closed for several other reasons which have to do with finance. One particular hazard - even for firms that run well - is the unfriendly takeover of a joint stock company, leading to closing of a plant whose production the new owner wants to locate somewhere else.

One reason why companies are exposed to this kind of hazard may be their own dividend policy. Rather than distribute all company profit as dividend to shareholders the management wants to safeguard against the hazards of running the firm by using part of the profit (the more the better) to build up reserves. Since many shareholders are mainly interested in dividend, the effect is that there is a big difference between the net value of the company and the market value of its shares. This facilitates a takeover attack through the stock market. On the other hand, an action like that drives the market prices up, and if it is not successful, the attacker may suffer a great loss. One example from the outskirts of the financial world: a Norwegian holding company controlling the biggest insurance company in the land joined forces with a Danish company in an unfriendly takeover bid aimed at a Swedish company. The attack failed and the Norwegian company which had been solid, well established, with great power and some international prestige, found itself insolvent.

The general assumptions of how a market economy deals with risk is that the one who gets the gain in the case of success, suffers the loss in the case of failure, and the decision-maker has the responsibility. The classic image of the capitalist businessman characterizes him as one who invests his own money, makes his own decision and picks up profits or losses. This species is not extinct but threatened. In big business the equity of big public companies comes to a large extent from small savings, and the shareholders know little and understand less of how their capital is operated. It is operated by chosen leaders in the board of directors, usually a well-chosen group of persons belonging to a circle of founding fathers who nominate them to be elected by shareholders who don't know one another but know the nominated persons as highly respected citizens. The board members are informed about current business and presumably understand more than the common shareholder. But the real leader is the managing director or a "working chairman" of the board. All these are well-connected through common interests, good fellowship, family relations and so on. The "managerial revolution" has made them the ruling class in business, and they have looked after themselves well and provide a security net for whoever gets into trouble. Their ideology states that leadership is leadership everywhere and a profession in itself. This is a modernized version of the feudal concept, surviving in England, that, if you are of good family, went to the right school and the right university, you are cut out to be a leader whatever your degree is. You may graduate in oriental languages and become a railway director. The modern version may be less aristocratic and more meritocratic in its entrance screening, but once you are admitted you are there.

The managerial salary is high because of your expertise, even higher because you are a leader, even higher because of your responsibility, even higher because of the risk that, if you lose your job, it is difficult to find an equivalent job. Finally, if it should come to the worst, your contract secures you a golden handshake for compensation. Should it really happen, the brotherhood will in addition see that you get a job worthy of your standing.

In the case of the Norwegian insurance company, placed under government administration, many employees, with no responsibility for the disastrous decisions, were fired, in many cases facing unemployment for years, maybe for ever. Shareholders lost their capital. The managing director was sacked with a contractual golden handshake amounting to several annual salaries in his sack. This was in no way a unique case. Such things happen with increasing frequency.

The self-protection of the managerial class against risk tends to make them more audacious in taking risks on behalf of the company and thus creates more risk in the world. The increased risk is carried by clerks, workers and shareholders.

Some anti-risk measures fail to serve their purpose. One source of uncertainty in the business world is change in currency exchange rates. The consequences of a system of rigid exchange rates were dramatically demonstrated in 1992. The struggle of the financial authorities to maintain the value of the national currency may involve otherwise unjustified dramatic increases in interest rates, such as to 500 % twice in Sweden in September 1992, very painful for industry and the common man. And the doubts that the effort will succeed increases the general feeling of uncertainty. The fear of a sudden devaluation or a drastic increase in the interest rate may be worse to live with than a continuous process of small day to day adjustments that we would have if we had a (free) market economy even in this field.

These are just a few examples of elements of economic risk which have admittedly been with us for a long time but which are more and more dominant in the general picture of our life.

Add to this the transition of more than a billion people who lived or still live (China) under the rule of a communist party from command economy (socialism) to market economy (capitalism). The managers in this area have no training or theoretical background in managing risk. The common man meets (or will meet) for the first time the risk of lasting unemployment.

We live in a world of uncertainty, we always have and, as indicated above, we do so to an increasing extent. Forces are at play to reduce risk, but the struggle to reduce one's own risk may not only shift risk to others but even create risk.

The above examples were concerned with economic risk, but we have corresponding phenomena where human values like life and death are concerned. One real life example: a group of soldiers were carrying out a manoeuver to undertake an operation which, due to unexpected weather conditions, might be too dangerous. Shortly before the arrival of an expert commission was due, the soldiers were sent out and more than a dozen died. Their commander was very sorry and shared his sorrow with the TV public. He solemnly declared that he took the responsibility. And that was the end of the history of his responsibility. Not quite. His courage and dexterity in handling responsibility was appreciated: he was headhunted into big business as a leader, with a salary high above what the army could offer.

When the matter of abolishing the death penalty was under debate in the British parliament, what probably impressed the house most was a former minister who as such had decided against a reprieve for a man condemned to death for having murdered his wife and daughter. After the man was hanged, new evidence showed that he was not guilty as charged.

In law, medicine and in many other cases decisions are taken with uncertainty about the facts. This concerns not only professionals, but also persons making decisions of a private nature, like Cathy. Theory of decisions under risk has its place in all social sciences and also in technology and natural science. The rules of statistical hypothesis testing belong conceptually to decision theory. The overweight of probability required to reject a hypothesis reflects the fear that research efforts will enter a blind alley by a too hasty decision.

In this world of uncertainty, of fear that a decision may be fatal and hope that it will lead to great success, those in a position to analyze scientifically the problems of uncertainty have a duty to offer assistance. But there is no use for demonstrations of proficiency in deducing from dogmas backed by academic authority. A theory is only scientific if it can be exposed to empirical testing. It is prescriptive only if it produces unambiguous advice of how to act in accordance with the preferences of those for which decisions are made.

Unfortunately, what has been presented as decision theory was for a long time exercises in reconfirming the beliefs of academic authorities. It was difficult to get papers published that presented what is the essence of scientific research: critical investigation into the logical and empirical validity of received theory. Even Allais, the later Nobel laureate, had difficulties. His Econometrica article (1953) was published after long discussions and with a kind of reservation: it was published "on his own responsibility.". From the end of the sixties a new tendency developed which gave an outlet for suppressed scientific contributions in this field. The organizational framework around this development was provided by the journal Theory and Decision, chief editor during a decisive period: Werner Leinfellner, later Bertrand Munier; Theory and Decision Library, chief editor Werner Leinfellner, also inspiring and playing a central leading role as president of the committee organizing a series of conferences starting in 1982 for the purpose of providing a free forum for discussion of "Foundations of Utility and Risk Theory," FUR, held bi-annually in or near Oslo, Venice, Aix en Provence, Budapest, Duke University, Paris and at The Norwegian School of Management.

Prof. Dr. Ole Hagen
Norwegian School of Management
P.O. Box 580, 1301 Sandvika, Norway

REFERENCES

Allais, Maurice [1952, 1979a], "Fondements d'une théorie positive des choix comportant un risque et critique des postulats et axiomes de l'Ecole Americaine", *Colloques Internationaux du Centre*

National de la Recherche XL, Paris 1953. Also in: Allais and Hagen (eds.) 1979, qq.v.: English translation.
Allais, Maurice [1953], "Le comportement de l'homme devant le risque: critique des postulats et axiomes de l'ecole americaine", *Econometrica 21*, pp. 503-546.
Allais, Maurice [1979b], "The so-called Allais paradox and rational decisions under uncertainty", in: Allais and Hagen (eds.) 1979.
Allais, Maurice [1983], "The foundations of the theory of utility and risk", in: Hagen and Wenstop (eds.) 1984.
Allais, Maurice [1985a], "Three theorems on the theory of cardinal utility and random choice", in: Eberlein and Berghel (eds.) 1987.
Allais, Maurice [1985b], "The Allais paradox", in: *The New Palgrave, A Dictionary of Economics Vol 1*, Macmillan, London 1987.
Allais, Maurice [1986], "The general theory of random choices in relation to the invariant cardinal utility function and the specific probability function. The (U,O) model", in: Munier (ed), *Risk, Decision and Rationality*, Reidel, Dordrecht 1987.
Allais, Maurice [1988], *Scientific Papers on Risk and Utility Theory - Theory, Experience, and Applications*. Forthcoming, Reidel, Ch. XVIII, "Cardinal utility - history, empirical findings, and applications."
Allais, Maurice and Hagen, Ole (eds.) [1979], *Expected Utility Hypotheses and the Allais Paradox*, Reidel, Dordrecht.
Allais, Maurice and Hagen, Ole (eds.) [1994], *Cardinal Utility*, Kluwer, Dordrecht.
Baumol, W.J. [1951], "The Neumann-Morgenstem utility index: an ordinalist view", *Journal of Political Economy 59*, pp. 61-66.
Baumol, W.J.: [1958], "The cardinal utility which is ordinal", *Economic Journal 68*, pp. 665-672.
Bemoulli, D. [1738], "Specimen theoriae novae de mensura sortis", *Commentarii Academiae Scientiarum Imperialis Petropolitanae 5*, pp. 175-205. (Exposition of a new theory on the measurement of risk, *Econometrica 22*, pp. 23-26).
Bitz, Michael and Rogusch, Michael [1976], "Risiko-Nutzen, Geld-Nutzen und Risikoeinstellung. Zur Diskussion um das Bemoulli-Prinzip", *Zeitschrift für Betriebswirtschaft 46*, p. 12.
Blasche, H. and Dorfner, E. [1987], "Selection by proxy: A model for the simplification of decision under risk and under uncertainty", *Theory and Decision 23(3)*, pp. 283-300.
Borch, Karl [1966], "A utility function derived from a survival game", *Management Science 12*, pp. 287-295.
Borch, Karl [1968], "Decision rules depending on the probability of ruin", *Oxford Economic Papers 20(1)*, pp. 1-10.
Chikan, A., Kindler, J., and Kiss, I. (eds.) [1991], *Proceedings of the 4th FUR Conference*, Kluwer, Dordrecht.
Daboni et al. (eds.) [1985], *Recent Developments in the Foundations of Utility and Risk Theory*, Reidel, Dordrecht.
Dahlbäck, Olof [1990], "An experimental analysis of risk taking", *Theory and Decision 29(3)*, pp. 183-202.
Geweke, John (ed.) [1992], *Decision Making under Risk and Uncertainty: New Models and Empirical Findings*, Kluwer, Dordrecht.
Hagen, Ole [1972], "A new axiomatisation of utility under risk. Teorie a Metoda", *Reprint 1987/4*, Norw., School of Management.
Hagen, Ole [1979], "Towards a positive theory of decisions under risk", in: Allais and Hagen (eds.) qq.v.
Hagen, Ole and Wenstop, Fred [1984], *Progress in Utility and Risk Theory*, Reidel, Dordrecht.
Hagen, Ole [1988], "Expected Utility Theory and Ordinalism: A Political Marriage", in: Munier Bertrand R. (ed), *Risk, Decision and Rationality*, Reidel, pp. 209-220.
Hagen, Ole [1992], "A Descriptive Model and a Decision Technique", in: John Geweke (ed.), *Decision Making under Risk and Uncertainty*. Kluwer Academic Publishers.
Hagen, Ole [1991], "Expected Utility Theory - The Confirmation that Backfires", in: Chikán A. et al. (eds.), *Progress in Decision, Utility and Risk Theory*, Kluwer, pp. 105-115.
Hagen, Ole [1988], "Expected Utility Theory and Ordinalism: A Political Marriage", in: Munier Bertrand R. (ed.), *Risk, Decision and Rationality*, Reidel, pp. 209-220.

Kami, Edi and Schmeidler, David [1986], "Self-preservation as a foundation of rational behavior under risk", *Journal of Economic Behavior and Organisation 7*, pp. 71-81.
Krelle, K. [1968], *Präferenz- und Entscheidungstheorie*, J. C. B. Mohr, Tübingen.
Krelle, K. [1984], "Remarks to Professor Allais' contributions to the theory of expected utility and related subjects", in: Hagen and Wenstop (eds.) qq.v.
Krzysztofowicz, Roman [1994], "Filtering risk effect in standard-gamble utility measurement", in: Allais and Hagen (eds.), *Cardinal Utility* qq.v.
Loomes, G. and Sugden, R. [1985], "Regret theory: An alternative theory of rational choice under uncertainty", *Economic Journal 92*, pp. 805-824.
Loomes, G. and Sugden, R., "Some implications of a more general form of regret theory", available from authors.
Lopes, Lola [1988], "Economics as a psychology: A cognitive assay of the French and American schools of risk theory", in: Munier qq.v.
MacCrimmon, K. R. and Larsson, S. [1979], "Utility theory: Axioms versus 'paradoxes'", in: Allais and Hagen (eds.) qq.v.
Machina, Mark J. [1983], "Generalized expected utility analysis and the nature of observed violations of the independence axiom", in: Stigum and Wenstop (eds.) qq.v.
Menger, Karl [1931], "Das Unsicherheitsmoment in der Wertlehre", *Zeitschrift für Nationalökonomie*.
Morgenstem, Oscar [1979], "Some Reflections on Utility", Allais and Hagen (eds.) qq.v.
Munier, Bertrand R. (ed.) [1988], *Risk, Decision and Rationality*, Reidel, Dordrecht.
Neumann, John von and Morgenstem Oscar [1953], *Theory of Games and Economic Behaviour*, Princeton University Press. Third Edition.
Pareto, Vilfredo [1916], *Trattao di Sociologia Generale*, Barbera Firenze.
Pope, R. [1985], "Timing contradictions in Von Neumann and Morgenstem's axioms and in Savage's 'sure-thing' proof", *Theory and Decision 18*, pp. 229-262.
Schoemaker, P. [1982], "The expected utility model: Its variants, purposes, evidence and limitations", *Journal of Economic Literature 20*, pp. 529-563.
Snow, Paul [1987], "Maximizing expected utility is a survival criterion", *Theory and Decision 22(2)*, pp. 143-154.
Stigum, B. P. and Wenstop, F. (eds.) [1983], *Foundations of Utility and Risk Theory with Applications*, Reidel, Dordrecht.
Stahl, I.: "Review" of Allais and Hagen (eds.) qq.v., *Scandinavian Journal of Economics 82*, pp. 413-417.
Sugden, Robert [1985], "Regret, recrimination and rationality", *Theory and Decision 19(1)*, pp. 77-100.
Sugden, Robert [1985], "New developments in the theory of choice under uncertainty", in: Hey, John D. and Lambert, P.J. (eds.) qq.v.

RATIONAL CHOICE THEORY: A CRITICAL LOOK AT ITS FOUNDATIONS

MARIO BUNGE

PREFACE

Marxism has been intellectually dead for a long time, and dictatorial state socialism is moribund. But this does not entail that the ideal of an egalitarian and democratic society is finished. Nor does it entail that capitalism and its ideologies are alive and well.

Firstly, nowhere is pure capitalism to be found. Indeed, every modern society since Bismarck and D'Israeli possesses some socialist features: this is what the welfare state is all about. Secondly, capitalism cannot prosper without exploiting and without directly or indirectly oppressing the Third World, where 77 percent of mankind lives, and which only gets 15 percent of the world income. Thirdly, capitalism has yet to find stable solutions to the following problems: war, unemployment, poverty, inflation, environmental degradation, and protection of the natural resources. Therefore it would be foolish to substitute market worship for Marx worship.

As the Pope said only recently, we in the West should not adopt a triumphalist attitude, for there are many things wrong among us. Rather, we should search for an alternative social order. However, in order to know exactly what we want and how we wish (and may and ought) to attain it, we should start by performing a critical analysis of the extant social sciences and political ideologies. In particular, we should note where science ends and ideology or even quackery begins.

The present paper analyzes the foundations of one of the most interesting schools of thought in contemporary social science, namely rational choice theory. Many economists, sociologists and politologists believe that this theory is no less than the vanguard of social sience. Here we ask whether the theory is clear, true, relevant and free from ideology.

INTRODUCTION

Rational choice theory assumes that (a) choice and exchange constitute the hub of social life, and (b) all choice is motivated by self-interest and, more precisely, guided by the principle of economic "rationality", or maximization of utility. Rational choice models have been mushrooming over the past half century in all of the so-called human sciences, from psychology to history.

Rational choice theory may be regarded as the culmination of the rationalist view of man held by such diverse thinkers as Thomas Hobbes and

J. Götschl (ed.), Revolutionary Changes in Understanding Man and Society, 211–227.

Baruch Spinoza, the French Enlightenment philosophers and the English utilitarians, as well as Auguste Comte, Karl Marx, and Max Weber. All of these thinkers stressed the role of interests, equated modernity with rationality, and rejected the Romantic as well as the religious models of man.

However, there are important differences between contemporary rational choice theory and its precursors. While the latter held a broad concept of rationality, rational choice theory uses the narrow instrumental concept of economic "rationality" or utility maximization formulated by the marginalist economists in the 1870s. Further, it disregards the problem of the rationality of goals (Weber's Wertrationalität), as well as that of the morality of both goals and means. And, whereas Comte, Marx and Weber thought in terms of a historical rationalization process in all domains of belief and action, rational choice theory takes it for granted that preferences are constant and that utility maximizing behavior is in human nature: It is an ahistorical view.

Rational choice theory can be criticized for being rational or for not being rational enough. The former kind of criticism is based on the writings of Schütz, Goffman, Geertz, Garfinkel, and Derrida, who in turn drew their inspiration from two of the main irrationalist philosophical schools of this century, phenomenology and existentialism. A closely related criticism of rational choice theory is the antimodernist "rhetoric of the human sciences", which claims that doing science is nothing but engaging in conversation and producing persuasive metaphors rather than maximally true data or hypotheses.

I am not interested in this kind of criticism of rational choice theory, except insofar as it occasionally makes a valid point, such as that real human beings are seldom fully "rational" - i.e., selfish and calculating - and that even mathematical economists make use of metaphors. My dissatisfaction with rational choice theory is that it is not rational enough, for involving an unrealisitic model of man and for skirting empirical tests. I will criticize it not for being scientistic but for not being scientific enough.

Believers in the power of reason are expected to build, apply, teach, or at least admire rational views of human behavior - behavior which is itself not always rational in any of the many senses of this ambiguous word. But it stands to (scientific) reason that this expectation is only justified if such views are supported by empirical evidence. In other words, rationalism is not enough: it must be combined with realism, i.e. a concern for the matching of theory with facts. Moreover, realism may be regarded as a condition of integral rationality, for it is irrational, both epistemically and instrumentally, to press for untestable or untested theories about any domain of reality.

This caution, unnecessary in the case of a mature science, is called for in the case of immature research fields such as sociology, economics and politology. Indeed, in these fields theory is often disjoined from data, while empirical research is frequently conducted in a theoretical vacuum. Worse, the philosophy underlying most social studies is either classical rationalism (reason is all-powerful) or classical empiricism (empirical data are the alpha and the omega of research). Rational choice theory illustrates the former research strategy, whereas the trial and error search for statistical correlations exemplifies the latter. In the following, I intend to substantiate the first conjunct of the preceding claim.

Being a rationalist of sorts - more precisely, a ratioempiricist - my view of rational choice theories is sad rather than jaundiced. Though an enthusiast of reason, I believe that rational choice theory has failed abysmally, and it saddens me that this failure has brought discredit upon the very enterprise of serious theorizing in the field of social studies. By the same token this discredit has brought aid and comfort to such foes of rationality as the symbolic interactionists, phenomenologists, ethnomethodologists, rhetoricians, and hermeneuticists. Therefore the following criticisms are intended as a contribution to making way for a more realistic and therefore more effective use of reason in the field of social studies.

I. BASIC PREMISES OF RATIONAL CHOICE THEORY

There is a large variety of rational choice theories, such as utility theory, decision theory, and game theory, as well as their specializations or models, such as neoclassical microeconomics, the Prisoner's Dilemma model of nuclear deterrence, Olson's model of collective action, Becker's model of the marriage market, and Coleman's "linear system of action". Let us begin by ferreting out the assumptions common to all such theories.

A rational choice theory or model of a domain of individual or social behavior is one that rests on the following axioms:

A1 Rationality Postulate: all agents are rational, in the sense that they act so as to maximize their (raw or expected) utilities.

A2 Ontological Individualism Postulate: all social groups are nothing but collections of individuals.

A3 Methodological Individualism Postulate: the properties and changes of any social group are understandable by studying the behavior of the group members.

These assumptions entail the following immediate consequences:

C1 Habit, compulsion, compassion, impulse, ideology and moral commitment play no role in determining individual behavior.

C2 Social groups have no emergent or systemic properties, i. e., they have only resultant properties.

C3 Social science has no need for irreducibly collective concepts such as those of social structure and social cohesion.
C4 Social science is reducible to a study of the utility maximization behavior of individuals.

The first three consequences are nothing but negative restatements of the corresponding first three axioms. The fourth is a corollary of the latter.

My first complaint is that the occurrence of the word 'rational' in rational choice theory is unfortunate for two reasons. First, the word in question is notoriously polysemic: it designates at least twelve different concepts. In fact, we should distinguish the following concepts of rationality. (i) Semantical: clarity or minimal fuzziness; (ii) logical: internal consistency; (iii) dialectical: inferential validity; (iv) erotetic: posing only problems that make sense in some context; (v) methodological: questioning and justifying assumptions and methods; (vi) epistemological: requiring empirical support and compatibility with the bulk of the background knowledge; (vii) ontological: admitting only ontological assumptions compatible with factual science; (viii) valuational: choosing only attainable goals; (ix) prohairetic: observing the transitivity of the preference relation; (x) moral: caring for individual and social welfare (xi) practical or instrumental: choosing or devising, in the light of the best available information, the means most likely to help attain the given goals; (xii) economic: self-interest. We shall group the first six rationality concepts under the rubric of conceptual rationality, and the last six under substantive rationality.

Second, it may be argued that utility maximization is often "irrational" - namely, whenever it is accompanied by shortsightedness, waste of human or natural resources, exploitation, oppression, war, or destructive competition - which may end up in self-destruction. Third and consequently, the use of 'rationalitiy' in that particular sense is persuasive or ideological rather than either descriptive or normative. It may serve as a disguise for self-interest. (I am rational, you use people, he is a shameless exploiter.)

This criticism does not involve a rejection of the rational approach to social facts. It does not even involve a denial that rational decision-making does play a role in human behavior, particularly in modern society - as rightly noted by Max Weber. I am only criticizing what I regard as a misappropriation of the ambiguous word 'rationality', as well as the thesis that self-serving calculation is the only or even the major guide of human action.

My second criticism is far deeper and, I hope, more interesting. It is that rational choice theory is far too ambitious. It claims to explain everything social in terms of just three assumptions that would hold for all individuals in all social groups in every historical period. But a Theory of Everything does not explain anything in particular. It may belong in logic or in ontology but it is certainly out of place in factual (or empirical) science. In particular, a Theory of Everything Social cannot account for both the band and the family, the small farm and agribusiness, the small workshop and the giant industrial firm, barter and commerce, marriage and divorce, education and politics, competition and cooperation, committee decision and social welfare judgment, war and peace, progress and decline, and so on and so forth. No wonder then that rational choice models are either far removed from reality or, when they do capture particular traits of real systems or processes, they do so by considerably enriching in an underhand manner the postulate system expounded above - e.g., by assuming that the individual agents interact via the market, which in turn is not analyzed in individualistic terms.

My third criticism is that, for being radically reductionistic (in particular individualistic), rational choice theory fails to account for the systemic constraints on individual behavior - or, in other words, for the effect of structure on agency. In short, the assumption that, at least in certain societies, the agents are completely free to choose, ignores the compulsions and constraints of various kinds that shape individual action. The theory overlooks the fact that one and the same individual acts differently in different social systems, just as a molecule in a liquid body behaves differently at the bottom and at the surface. True, the rational choice theorist is often apt to invoke changing "situations" or "circumstances", but - as noted before - he treats these as black boxes, and so in a holistic manner inconsistent with his individualistic approach.

So much for general critical remarks. In what follows I shall spell these out and shall examine a number of particularly interesting, albeit highly controversial, rational choice models. I shall start with the Rationality Postulate; but since this involves the notion of utility, we must first examine the latter.

II. UTILITY

The concept of utility, or subjective value, plays the same central role in neoclassical economics and the other rational choice theories as that of objective value in classical economic theory. Although a popular dictionary of economics asserts confidently that "The concept of utility is not contentious", experts in utility theory have often expressed qualms about it, as can be seen by perusing the *Theory and Decision* journal.

I submit that the general utility concept has two fatal flaws. The first is that only its existence has been proved. It is an undefined function - unlike, say, the sine or the logarithmic functions. While such generality is desirable in abstract mathematics, it is a severe shortcoming in any chapter of applied mathematics, where computation is of the essence. The second flaw is that, being subjective, utility is hardly measurable or even subject to interpersonal comparison. In other words, the values of the utility functions are not assigned either rationally (by calculation) or empirically (by measurement). This is why in social studies they are usually assigned intuitively - hardly a triumph of rationality.

In fact, the standard von Neumann-Morgenstern axioms only imply the existence of (infinitely many) utility functions, without defining any of them. And the utility functions occurring in microeconomics are only slightly more specific. In fact, from the beginning of the so-called marginalist revolution, the utility functions have been subject to just two qualitative conditions. These are that any utility function must be monotonically increasing and decelerated. (I.e., $du/dx > 0$ and $d^2u/dx^2 < 0$.) In other words, the "law" of decreasing marginal utility (du/dx) with increasing quantity (x) is assumed more or less tacitly.

Such lack of specificity is similar to that of the general distance function. Indeed, in the general theory of metric spaces one only demands that the distance function d complies with three conditions: for all elements ("points") x, y, z in an arbitrary set S (the space), $d(x,x)=0, d(x,y)=d(y,x)$, and $d(x,y)+d(y,z) \geq d(x,z)$. These conditions define any of infinitely many distance functions in spaces of arbitrary dimensionality. By the same token they do not specify any particular space. In order to compute or measure the distance between two points in a particular metric space one needs a specific distance function, such as the one defined by the Pythagorean theorem. A physicist who were to limit himself to asserting that physical space (or spacetime) is a metric space, without committing himself to the precise metric (or distance function, or line element), would be laughed out of the laboratory. On the other hand, social scientists can get away with similar generalities concerning utility.

To put it another way, a function u about which one only knows that it increases monotonically, though with a decreasing slope, is an unknown. Hence to claim that one understands an action by saying that its actor a is attempting to maximize his u, amounts to saying that a behaves the way he does because a has a certain occult quality. This is a case of explaining the obscure by the more obscure. It is theology, not science.

It may be rejoined that, unlike the grand theories in traditional social studies, some of the rational choice models are chock full of numbers. True, but these values happen to be taken out of a hat. It does not help assigning the utility functions definite values (in particular maxima) if these values are posited rather than either calculated or measured. And this is precisely what most rational choice models do, namely, to postulate the basic utilities instead of either computing or measuring them.

The subjectivity of (both cardinal and ordinal) utilities makes it impossible to state, let alone prove, that a given person or group is objectively better or worse off than another, or that one business deal or political manoeuvre is objectively preferable to another. Like beauty, subjective utility and its changes are in the eyes of the beholder.

One consequence of this is that there would be no such thing as objective inequality or social injustice - a convenient result for the opponents of social reform. A second consequence is that since subjective utility statements are not rigorously testable, the theories in which they play a key role - such as neoclassical economics - are in the same boat as philosophical aesthetics (as distinct from psychological or experimental aesthetics). That is, they are nonscientific. No wonder that the more methodologically sophisticated social scientists treat cardinal utility in a cavalier fashion or dispense with it altogether.

What holds for subjecitve utility concept holds, *a fortiori*, for that of expected utility, which occurs in the formulation of the Rationality Principle and involves the probability concept. The probability in question is not well-defined either - except, again, in the case of games of chance. In fact, in most rational choice models probabilities are assigned in the same arbitrary manner as utilities. More precisely, they are usually taken to be *subjective* or *personal* probabilities.

The adoption of subjective or personal probability values is of course consonant with the Bayesian school of mathematical statistics defended by H. Jeffreys, B. de Finetti, L. J. Savage, I. J. Good and a few others, as well as with Bayesian decision theory. But it is inconsistent with the way probabilities are assigned in mature disciplines such as statistical mechanics, quantum mechanics, genetics, and even some branches of social science, such as social mobility theory. In these disciplines probabilities (or probability distributions) are treated as objective properties, like lengths and populations, neither of which is supposed to be estimated subjectively (except provisionally and subject to experimental checking).

Moreover, in science probabilities are only introduced when there is reason to believe that a chance process, such as random shuffling, is at work. No probability without objective randomness. Yet Milton Friedman assures us confidently that "individuals act as if they assigned personal probabilities to all possible events". The fictionist "as if" trick renders the statement untestable and thus frees the theorist from the burden of empirical testing - very convenient for the lover of free speculation. But it leaves the statistician baffled, for in the Bayesian context there is no room for the concepts of randomness, randomization, and random sample.

Finally, a major problem with most utility theorists is that they overlook adverse empirical evidence, such as that produced by the survey conducted in 1952 by Nobel laureate Maurice Allais. This survey, involving a hundred subjects well-versed in probability, showed that rational people are not bent on maximizing anything except security. Additional empirical studies have exhibited a moral dimension absent from rational choice theory. For example, Kahneman, Knetsch and Thaler found that people tend to behave fairly even at some cost to themselves. Many other economic psychologists, by exploring the variety of human motivations, have refuted the myth that man is above all a utility maximizer. By refusing to face such adverse empirical evidence and sticking instead to 19th century armchair psychology, rational choice theorists behave like true believers, not like rational agents.

III. RATIONALITY

As noted in Section I, the word 'rationality' designates at least a dozen different concepts, only one of which has been singled out by most rational choice theorists. Consequently, one should be able to conceive of at least twelve "rationality postulates" in addition to the Rationality Postulate called A1 in Section I. For example, in collective choice theory the rationality postulate boils down to respecting the transitivity of every preference relation. And in action theory an action is called 'rational' just in case it is conceived of and executed in the light of the best available information.

It is instructive to place Al alongside four alternatives found in the social studies literature. They are

- Rl "Agents always act in a manner appropriate to the situation in which they find themselves" (Popper).
- R2 Principle of Instrumental (or Functional) Rationality: agents always adopt the means most likely to produce the desired results.
- R3 = Al Principle of (Economic) Rationality: agents always act so as to maximize their expected utilities.
- R4 Principle of Least Effort: agents always choose the least expensive means to attain their goals.
- R5 Principle of Subjective Rationality: agents always act on their beliefs about the situation in which they find themselves, as well as on their beliefs about the most suitable means and the possible consequences that their actions may have for themselves and others, and aim for the consequences they judge best.

Rl is so vague that it might be said to hold for an electron in an external field as well as for a human being. Indeed, both act according to their circumstances. Moreover Rl is hardly testable, if only because the very concept of "appropriate behavior", on which it hinges, is fuzzy. Not being rigorously testable, Rl is neither true nor false.

The principle R2 of instrumental rationality is false. Indeed, comparatively few people are intelligent and well informed enough, as well as sufficiently resourceful and strong, to be able to employ the best available means, much less to do so always. If R2 were true, men would seldom if ever resort to war to settle their differences. If R2 were true, there would be no market for shoddy products such as "lemons" (bad cars); there is such a market because the buyer ignores something that the seller knows. At best R2 could be regarded as a definition or else a norm or ideal of instrumental rationality. More on this matter at the end of this Section.

The principle R3 of economic rationality, i.e. our old acquaintance Al, is false because it implies the weaker principle R2, which is false, as we just saw. However, we shall see below that R3 may be regarded as irrefutable rather than false.

Morgenstern would have objected that R3 is neither true nor false, for being a norm or rule of behavior rather than a descriptive statement. An obvious rejoinder is that R3 is a poor guide to action because it does not specify whether one should seek to maximize short, medium or long term utility. If these three possibilities are conflated, and particularly if short-term self-interest is allowed to prevail, one may end up by falling into "social traps". Familiar cases of social traps are overexploitation (in particular labor and colonial exploitation, overgrazing and overcultivation), overindustrialization, overurbanization, overautomobilization, militarism, and wars. In conclusion, the maximizing behavior postulate R3 is worse than either untestable or false: its observance is the ultimate cause of the global disasters humankind is facing. Some rationality!

R4, the principle of cost minimization, is false for the same reason that R2 is.

R5 is my own version of Boudon's principle of subjective rationality. It is so weak as to be nearly tautological, immune to refutation. On balance, three of the principles are false and two, to wit Rl and R5, are hardly testable. Indeed, suppose someone behaved in a manner that an observer, such as a social scientist, would ordinarily regard as irrational - e.g., from habit, on impulse, from a love of gambling, under compulsion, out of commitment to an absurd doctrine, without sufficient information, or in a highly original way. The agent could argue that, under the circumstances, his action was rational after all, particularly since he knows best what is in his own interest.

Yet, it would be irrational to give up rationality in social studies and adopt instead an irrationalist or Romantic approach. What we need in the social sciences and technologies are (a) a rational (and empirical) approach to human behavior of all kinds, rational and nonrational, (b) a sufficiently clear concept of instrumental rationality (Weber's Zweckrationalität) and (c) a normative or prescriptive, rather than descriptive, principle of instrumental rationality. The following are offered for discussion.

Definition of "instrumental rationality": An action is instrumentally rational if and only if it is likely to produce the desired outcome with the least effort.

Norm of "instrumental rationality": A rational individual engages deliberately in an action M at a certain time if and only if (a) M is a means to his goal G, as suggested by the best available information, (b) G has priority over other goals of his at that time, and (c) the cost of M, added to that of the undesirable effects accompanying G, is smaller than that of any other means known to him, as well as smaller than the value of G.

This is a prescription rather than a description, because it is restricted to instrumentally rational individuals. (On the other hand it is not restricted to humans. It may well apply to subhumans as well as to humans.) It is also restricted to instrumental rationality: the norm does not touch on alternative concepts of rationality, let alone on overall rationality. In particular, no mention is made of maximizing any utilities.

If an individual makes the (practically) wrong choice of means, he is said to behave in an instrumentally, objectively irrational way, not in a subjectively rational way. But the means he chooses, though instrumentally irrational, may happen to be rational in some other ways, e.g., morally. I.e., he may have to trade rationality in some regard for irrationality in another. Finally, the genuinely rational agent is supposed to weigh his various values and goals at any given moment and to refrain from sacrificing everything to the overriding value or goal of the moment. Total commitment to a single value or goal, other than individual and social welfare, is self-defeating in the long run because life is many-sided.

In sum, neither of the various rationality principles found in the literature has a descriptive value. Simpler, real people do not always act rationally, in any of the senses of 'rational'. However, the principle R2 of instrumental rationality is valuable when reworded in the normative mode, i.e. as a guide for action. Moreover it is useful even in the original declarative mode, namely as a methodological or heuristic tool. This is in fact how Max Weber understood it. He regarded it as an idealization allowing one to explain certain human actions as outcomes of (instrumentally) rational decisions, and others as departures from rationality originating in irrationalities of all kinds, such as effects and mistakes, not to mention natural and social forces beyond the agent's control. (Mechanical analogue: account for any departures from either static equilibrium or rectilinear uniform motion in terms of forces or constraints.)

In short, we uphold conceptual rationality, which is an ingredient of the scientific approach, and the Norm of Instrumental Rationality, which is a guide to efficient action. But we cannot accept any of the so-called rationality principles that occur in rational choice theory. Indeed, we reject Rl for being vague and therefore untestable, R2, R3 and R4 for being false, and R5 for being nearly tautological.

IV. INDIVIDUALISM

Rational choice models are ontologically as well as methodologically individualistic. That is, they focus on individuals and either deny the existence of social bonds and social systems, or assert these to be reducible to individuals and their actions.

Ironically, there are few if any consistent upholders of ontological individualism or social atomism. Even such radical ontological individualists as Weber, Simmel and Homans have occasionally admitted the existence of social systems and have always resorted to such unanalyzed wholes as "the market", "the State" or "the situation" - without realizing the inconsistency.

The ontological individualist views society by analogy with a low density gas, where the individual agents are the analogues of the molecules, and the former's utility functions are the counterparts of the laws of motion. As in the elementary kinetic theory of gases, the individuals are assumed to act independently of one another, being only constrained by the container (or its analogue, the market). Incidentally, neither the gas container nor its social analogue is ever analyzed as a collection of individuals: both are handled as wholes - a strategy that violates methodological individualism.

But social systems, from the family and the firm to the transnational corporation and the State, are more like lumps of condensed matter than like gases. And the theories of condensed matter bear little resemblance to the theory of gases - which is as it should be, for condensation is accompanied by a qualitative change. In these theories interaction is of the essence: it is what makes a body fluid or solid, just as face-to-face interaction turns an amorphous social group into a social system. This is why few social scientists (unlike social philosophers) practice ontological individualism.

While there are few if any consistent ontological individualists, there are plenty of methodological individualists, i.e., students of society who claim that the understanding of social facts requires the investigation of the beliefs, intentions and actions of all the individuals concerned. However, methodological individualism, hence rational choice theorizing, comes in at least two strengths: strong and moderate. The strong version holds that a single theory, containing concepts referring exclusively to individuals, suffices to explain all kinds of social behavior and social system. The self-styled "economic imperialists", who claim that neoclassical microeconomics suffices to explain everything social, from marriage to war, belong to this party.

On the other hand moderate individualism holds that the universal premises concerning individuals, though necessary, are insufficient to account for social life. It asserts that different kinds of social fact call for different models sharing those universal premises, and that each model contains hypotheses concerning both the specific kind of interaction and the particular institutional framework. Moderate individualism comes close to systemism. Let us take a closer look at the two varieties.

The strong version of rational choice theorizing holds that all social groups - small and large, formal and informal, voluntary and involuntary - must be accounted for by a single overarching theory. This assumption can be seen as a consequence of methodological individualism conjoined with the hypothesis of the uniformity and constancy of human nature. Indeed, if all individuals are only led to act by self-interest, and behave regardless of the way other people behave, then all social groups are essentially the same, and consequently a single theory should fit them all. Moreover, since self-interest is the key concept of the paragon discipline, namely neoclassical microeconomics, all rational choice models must be variations of the latter - i.e., all of social science is ultimately reducible to microeconomics. This may still be a minority view, but it has been steadily gaining ground during the 1970s and 1980s.

There are at least three objections to this radical view. The first was mentioned in Section I, and it is summarized in the classical formula "Dictum de omni, dictum de nullo" - i.e., what is said about everything says nothing. (In semi-technical semantic jargon, sense or content is inversely proportional to extension or truth domain.) Put another way, an extremely general theory can only cover the features common to all the members of its reference class, hence it will miss all peculiarities. For example, it will fail to distinguish altruism from crime, churches from business firms, schools from armies, and so on. And, being unable to account for differences among individuals and for the variety of social interactions, systems, processes, and institutions, the theory is bound to be unrealistic, i.e., false.

A second objection to radical individualism is that social facts, and even individual behavior, cannot be accounted for without the help of certain "macroconcepts" that do not seem definable in terms of "microconcepts". (Obviously, a macroconcept is a concept referring to a macroentity, such as a business firm, and a microconcept one concerning a microentity, such as an individual person.) For one thing, behavior does not happen in a social vacuum but in a social matrix. For another, as Granovetter has shown, "most behavior is closely embedded in networks of interpersonal relations".

A consequence of the embeddedness of the individual in social systems of various kinds is that a genuinely rational decision maker will take his social environment into account. For example, a businessman cannot make rational choices unless he takes into account the global entities or processes referred to by such concepts as occur in the following laundry list: "policy of the firm", "state of the market", "organization", "technology", "product quality", "scarcity", "bottleneck", "economic conjuncture", "development", "business opportunity", "social order", "value added tax", "inflation rate", "prime rate of interest", "labor movement", and "political stability".

None of the concepts we have just listed is about individual persons and none of them is definable in terms of individual dispositions or activities. All of them transcend biology and psychology. They are peculiar to social science because they concern social systems embedded in even larger systems, and every system possesses (emergent) properties that its components lack. For instance, a country may be at war without every one of its inhabitants taking up arms. Another example: capitalism, socialism, and their various combinations are features of entire economies, not of individuals. A third example: parliamentary democracy can only be a property of a political system.

Emergence sets objective limits to microreduction, or the construal of a whole as the collection of its components. It cannot be otherwise if our ideas are to match the real world, which is one of changeable systems. Interestingly enough, this is the case with atomic physics as well as with social science. Indeed, every well-posed problem in quantum theory involves a formulation of the boundary conditions, which represent the environment of the thing in question in a global manner rather than in terms of "microconcepts" representing microentities.

The weak version of rational choice theorizing admits the need for subsidiary hypotheses concerning both the specific type of social group, and (though often in an underhand fashion) the institutional framework or social environment as well. Every rational choice model of this kind may then be regarded as the set of logical consequences of the union of the set G of general premises of "rational" behavior discussed in Section I, and a set of subsidiary or special assumptions describing the specific properties of the system or process in question. The role of G is thus similar to that of the general principles of evolutionary biology, which are insufficient to account for the differences between, say, bacteria and giraffes.

This weak version of rational choice theorizing is not open to the two objections we raised against the strong version. However, even this moderate version is unrealistic because G itself is unrealistic. Indeed, people act from custom, obligation or passion as well as on the strength of cold calculation. Moreover, not even calculated behavior can be perfect, and this for at least two reasons. Firstly, the information required by any realistic calculation is never complete. Secondly, the goal aimed at may not be within the reach of the agent, or it may be morally irrational.

Another unrealistic assumption of both the weak and the strong versions of rational choice theory is that of freedom of choice. Even a cursory examination of any of the major events in anyone's life suffices to refute this view. For example, we cannot choose whether or when to be born or to come of age: whether in an advanced or a backward country, in good times or bad. In hard times we cannot be choosy about jobs or lifestyles; we can seldom choose whether to live in a hovel or in an Alpine cottage; the education we get and the occupation we engage in depend crucially on the socioeconomic status of our family; the citizens of most nations, however democratic, have little if any say in choosing between peace and war - and so on and so forth. No doubt we ought to promote the cause of freedom (and the concomitant duties), but the idea that we can attain total freedom is at best an illusion, at worst an ideological bait. And if our choices cannot be totally free then they cannot be fully rational in any sense of this word.

In sum, neither ontological nor methodological individualism works. By ignoring the fact that individual action is embedded in social systems, they provide an "undersocialized" view of man and society, just as holism proposes an "oversocialized" picture. We should therefore steer clear from both extremes, as well as from the cryptoholism involved in what we have called individholism. The alternative to all three is systemism, for it accounts for both individual and system, or agency and structure. Indeed, systemism models every system, social or not, as a triple (Composition, Environment, Structure), so that it encompasses the valid features of its rivals.

V. CONCLUSION

Rational choice theorists have offered rather simple, hence attractive, explanations of hundreds of social facts of all kinds. All these explanations assume individualism, free choice, and maximizing behavior - the core hypotheses of mainstream economic theory and utilitarian ethics theory from Adam Smith and Jeremy Bentham on.

Although rational choice models have become increasingly sophisticated, nearly all of them share the following fatal flaws:

a) they do not define the (subjective) utility functions, which play a central role in them; hence, except for the rare bird, they are pseudomathematical despite their symbolic apparatus;
b) they contain the concept of subjective probability - which, being subjective, does not belong in science;

c) they involve the concept of subjective rationality, or rationality in the eye of the agent - which serves to save the Rationality Postulate from refutation;
d) they are static: in particular, they assume the fixity of options and the constancy of preferences (or utilities);
e) although they are supposedly strictly individualistic, they smuggle in unanalyzed holistic notions such as those of "the situation", "the market", and "the institutional framework" - whence they are individholist rater than consistently individualist;
f) they overlook the power of elites, governments, bureaucracies, corporations (in particular oligopolistic and transnational firms), networks of all kinds, lobbies, and other groups and systems, all of which distort preferences and restrict the freedom of choice, and justify Lafontaine's famous statement, "La raison du plus fort est toujours la meilleure";
g) they willfully ignore the planning that guides every modern social system, from the family and the retail store to the State and the transnational firm, and that constrains individual choice and action;
h) they only cover one type of social interaction, namely exchange, and even so they often do so via some unanalyzed whole, such as the market;
i) they omit all the sources of human action other than self-interested calculation - in particular they overlook habit and coercion, as well as passion and compassion;
j) by focusing on choices and means, they underplay motivations, goals und side effects;
k) they do not solve the problem of the micro-macro relations and, in particular, they do not account for the agency-structure relations or for the emergence and breakdown of social systems;
l) because of their excessive generality they have almost no predictive power;
m) for the same reason they are unrealistic: they are mostly jeux d'esprit, even though some of them have been sold to big business or to the military establishment;
n) being unrealistic, at times surrealistic, they are not suitable policy tools.

In sum, we have found rational choice theories inadequate. We have found them wanting not for being "trapped within reason" - the irrationalist reproach - but for being trapped within the individualist and utilitarian dogmas, for involving fuzzy basic notions and untestable key assumptions, for idealizing the free market, and for failing to match reality.

This is not to say that all rational choice theories are utterly useless. In fact, they can perform three services: (a) they may help hone the preliminary formulation of some problems, by forcing the student to distinguish the options, draw decision trees, and take costs and benefits into account; (b) some of them show that the pursuit of self-interest may lead to "social traps", i.e., collective disasters; (c) others may suggest experiments aiming at testing those very models, such as the experiments which have been conducted to check game-theoretic models; and (d) by failing to represent reality they may suggest the search for deeper and more complex theories - much as the failure of the flat earth hypothesis stimulated the elaboration of the round earth one. In sum, rational choice theories, if not taken too seriously, may serve heuristic functions. However, too many rational choice models serve only as intellectual gymnastics, for academic promotion, or to justify aggressive military strategies.

Why then does rational choice theory enjoy such enormous prestige? For the following reasons: (a) it is basically simple; (b) it claims to explain all human behavior; (c) it makes liberal use of symbols that occasionally designate well-defined mathematical concepts and always intimidate the non-mathematical reader; (d) it makes predictions - alas, not often correct ones; (e) it endows market worship with intellectual respectability; (f) it is often criticized for the wrong reasons, and (g) so far it has no serious rivals.

To sum up, rational choice theory is not a solid substantive theory of society. This failure can be traced back to the specific assumptions of the theory, not to the rational approach to social studies. Any serious study of society will be rational, but no realistic account of social facts will assume that people always act rationally, in particular selfishly.

The reader may feel, like my friendly critic Professor Werner Leinfellner, that my criticism is excessive: that I am throwing the baby out along with the bath water. My reaction is that there is no baby. In fact, rational choice theory was born in 1871 in the field of economics, so it is anything but young. Worse, it died long ago from mathematical anaemia, from deficiency in the enzymes required to digest the simplest social facts, and from lack of exposure to the sound and light and fury of the social weather.

To conclude, allow me to propose seven insolent, though possibly pertinent, maxims for the guidance of the perplexed student of social reality:

M1 Do not mistake cupidity for rationality.

M2 Being rational does not involve imputing rationality to everyone else.

M3 Regardless of one's admiration for rationality, one should never underrate the power of irrationality, especially one's own.

M4 Mathematics should be a servant, not the master, in the task of modeling reality.

M5 Get in touch with social reality once in a while. Meet people, mix with them, and join with them in common enterprises, so as to get a feel for social systems and their changeability, and to overcome the prejudice that everyone else is as selfish bastard.

M6 Avoid inbreeding. Do not just read other scholars' papers - also read the press and listen to what people have to say.

M7 When all else fails, try the scientific approach. It has been known to work.

Prof. Dr. Mario Bunge
Foundations and Philosophy of Science Unit
McGill University
3479 Peel Street
Montreal, PQ, Canada H3A 1W7

COMPLEXITY, THE CONCEPT OF UNCERTAINTY AND BOUNDED RATIONALITY OF MAN

BERTRAND MUNIER

Our Societies become more complex everyday, particularly in the economic sphere. This is due not only to technology, but also to organizational needs in every social cell, from government services down to the individual. Everyday even the ordinary person is becoming more aware of the necessary links between today's societies across the world, and feels that human behavior must be adapted to this new situation. This law of emerging "Complexity-Conscience" has long been known [Teilhard de Chardin, 1959]. One can also see, in our societies, that our decisions and actions are prepared and applied under a very strong *uncertainty.*

Herbert Simon turned the notion of *bounded rationality* into a basic ingredient of economic theory and management practice. Man has bounded combinatorial capacity as well as bounded memory and thus is significantly more limited in preparing his decisions and dealing with them than conventional economics had previously led us to believe. How human beings succeed in dealing with complexity and uncertainty in this world by means of a bounded rationality is thus a quite natural question.

Are today's societies becoming less and less well managed, as a result of some sort of impossibility theorem of the social sciences? If yes, is this development completely unavoidable? Or is Man in a situation to control society, particularly the economy, just as well as, or perhaps even better than in the past? If that is the case, how then does he solve the obvious opposition between, on the one hand, complexity and uncertainty, which he tries to escape, and on the other hand the limitation of his perceptive faculties, from which he needs to find a way out?

In order to answer these questions, one must first, clearly identify the concepts of complexity, uncertainty and bounded rationality, and second, explore the relationships between complexity and uncertainty. The organization of this essay is then straightforward. After an introductory section, devoted to an examination of the concepts, I explain in section 2 how *complexity implies limitations to human control of the economy.* In section 3, I try to show how bounded rationality allows Man to retrieve his ability to manage economic organizations, by making use of the complexity-uncertainty couple to overcome the already mentioned opposition between its two elements.

J. Götschl (ed.), Revolutionary Changes in Understanding Man and Society, 229–242.

1. INTRODUCTION OF THE CONCEPTS

1.1. Concept of Complexity

That we all live in a complex world is a proposition on which we easily can agree. In such a context, however, it is quite likely that we would draw no distinction between "complex" and "complicated". Nevertheless, I suggest that we draw a borderline between "complexity" and "complication".

A problem is complicated when the connections between the elements of the operating system are numerous and so sophisticated that much time, great effort and a lot of money is required before we can state that we have a reasonable understanding and an acceptable model of the phenomenon.

By contrast, a problem is complex, when it turns out to be impossible to come close to any acceptable working model [J.L. Le Moigne, 1977, or Santa Fe Institute]. There are logical reasons why our mind cannot produce any relevant models. As I shall argue, such situations are not uncommon in today's economy. In any case, I shall, in what follows, use the words "complex" and "complexity" in this sense of the counterintuitive and of the unperceivable.

1.2. Concept of bounded rationality

The traditional concept of rationality in economics is the one of maximizing of a utility function, i.e. of optimizing a *one dimensional* criterion, which supposedly works as a "synthesis" for our numerous grounds for judgement. When we unambiguously know the value taken by this function as a result of the selection of a specific decision (i.e. when we are in a so-called "certainty" situation), applying this concept of rationality entails no technical difficulty. When we, however, do not possess such knowledge, the application takes an entirely different turn: different events can occur which, according to the particular decision taken, entail respectively corresponding results. One then speaks of decision-making under risk or uncertainty. In this context, John von Neumann and Oskar Morgenstern proved in Princeton in 1944 that, under a specified set of axioms, the corresponding generalized concept of rationality is the maximization of expected utility. This usual view will be referred to here as the concept of "strong" rationality.

Herbert Simon took a strong stance against this view and argued that Man's cognitive capacity is so limited that maximizing expected utility is completely out of his reach. Indeed, such an attempt could even turn out to be dangerous, to the extent that we never are sure that *every* possible dimension of the problem has been considered: It is well-known that optimizing on an incompletely specified set can lead precisely toward the opposite direction with respect to the right solution. The opinion of Herbert Simon [Simon, 1955] is thus that we should be satisfied if we reach a given threshold of the utility function or of any equivalent evaluation device. This view is the first and most well-known version of the concept of "bounded rationality". It can also be characterized within the framework of the well-known I-D-S model, consisting of an Intelligence phase, a Design phase and a Selection phase.

The neo-classical school of Economics has considered this view as relatively secondary because it has been argued that bounded rationality is a side-product of the usual concept of strong rationality. Indeed, George Stigler elaborated his theory of "economic search" with such a perspective in mind. According to that model, the economic agent first accumulates enough information before he or she makes a decision. How does the housewife, for example, stop looking for more information on the possible qualities and prices of a commodity? She simply compares her aspiration level and expected utility: the latter quantity can be computed from the possible commodities and qualities and from the probability distribution of the corresponding prices which has emerged from the search. Such a rule appears, in fact, as a criterion of satisfaction: as soon as expected utility has reached the aspiration level, the housewife puts an end to her search and makes a decision.

Some authors [Laux, 1982, p. 54-55, e. g.] have gone a step further and regard bounded rationality as a reasonable complement of the usual theory of decision. They even argue that it is simply a special case of the usual optimizing rationality. It is quite simple, indeed, to introduce a similar but optimal rule into the theory of search. One only needs to introduce the costs of the search into the computation and to reformulate the rule in the following way: the housewife (to take again the same example) will put an end to the search when her expected increment in utility from any incremental search is not larger than the expected cost of it.

The next question is then: does Herbert Simon say effectively something more, and does bounded rationality mean something effectively different, with respect to the usual concept of strong rationality, if one takes the pain to underscore that the optimal allocation of scarce means must leave some space for a search of information prior to the choice between different courses of action?

The answer is yes. It is an entirely different concept. The example already used above can be rejected on two grounds:

- On empirical grounds: the necessary computations demand even larger cognitive capacities from the decision-maker than the traditional maximization problem [Simon, 1979]
- On a more specific, logical ground (which Herbert Simon has never really dealt with, although he made mention of it): the costs of the search, as well as other elements of information, which appear to be necessary to the application of search theory, are usually not available to the economic agent [Simon, 1955, p. 254]. *Thus, the economic agent will again have to search for elements of information.* But in this second order search, he or she will not have the necessary information anew, etc... This leads us to an infinite regression of evaluation which is perhaps solvable from the point of view of the observer, if a fixed point exists [Sidney Winter, 1975, p. 83, Mongin et Walliser, 1988], but which will never be solvable from the point of view of the economic agent himself, because the infinite sequence of the costs of the incremental searches will never be known to him [Mongin, 1984, p. 552-553]. Man will thus never be sure whether he has reached the optimal point.

In fact, Herbert Simon has pointed out another aspect of bounded rationality: our cognitive capacities being limited, it might be better to simply try avoiding catastrophic decisions, and to renounce looking for the optimal solution. With such a goal, all we need, far from any sophisticated algorithm which would designate to us "the best" solution, is a well-ordered *procedure* of searching for information and of dealing with it. This is the subtle distinction between the (bounded) *"procedural"* rationality and the (strong) "substantive" (or "*instrumental")* rationality. At this point, no doubt can remain, because no simple art of optimization can clearly match this fundamental concept of the *treatment of information* or of the *deliberative evaluation*. It is precisely here that we find the correct distinction between the usual concept of strong rationality and the concept of bounded rationality. Bounded rationality emerges from this view of decision-making issues as a *process* of reason more than as a *product* of reason.

The Simonian I-D-S model, which appears as a basic frame for this process, should nevertheless be complemented by the fact that institutions of our society help Man during the entire decision process, to the extent that they perform not only an effective function of coordination of behavioral strategies, but also - a fact much less frequently mentioned - of opinions and expectation formation. The market, for example, causes a relatively marked convergence among opinions: when, to stay with this example, a large majority of traders thinks that this or that currency will lose value on the market, a trader who thinks the contrary and puts his thinking into practice will be difficult to find (fortunately for this trader!). The extreme case of such situations has been described by Keynes in chapter twelve of the "General Theory... [1936] under the name of *convention*.

This complement to the theory of bounded rationality can easily be introduced in the Intelligence phase of the I-D-S model, but has been only little and not clearly dealt with by Herbert Simon [Simon, 1983, p. 380, Munier, 1991]. This is undoubtedly the big loophole in the Simonian theory.

This *concept of the deliberative evaluation* appears as particularly useful in the *prescriptive* applications of decision theory, which represent the ideal field for applying the Simonian model. In such applications, "procedural" rationality emerges as most useful, particularly in the Intelligence and Design phases: the best general is not the one who most precisely allocates his forces among different areas on the battlefield, but more often the one who can sketch a new, until then unperceived, possible strategy!

1.3. Concepts of Risk and Uncertainty

When different events capable of influencing the consequences of our decisions can occur, one says that we are in situations under risk or under uncertainty. The difference lies in our knowledge of the probabilities associated with the relevant events. If we know these probabilities unambiguously, the problem is said to be one of decision under risk. If we don't, it is said to be one of decision under uncertainty. F. Knight [Knight, 1921] is usually credited with this distinction (although there may be some dispute as to whether F. Knight did not have something quite different in mind). Savage [Savage, 1954] has criticized this distinction, but one cannot say that, in general, today's scientists have a single view on this issue.

The question of the nature of uncertainty has been much less dealt with than this last distinction. Thus, I would like to show, in the second part of this talk, how complexity can make it impossible for Man to control the economy and how this phenomenon can lead to new insights into the nature of uncertainty.

2. COMPLEXITY AS A LIMITATION TO HUMAN CONTROL OF THE ECONOMY

2.1. Nonlinear behavior

Modelling economic behavior has changed more in the last ten years than in the immediately preceding quarter of a century. In a *descriptive* perspective, in which the "strong" rationality of maximization may keep its value, one should emphasize the fact that the processing of probabilistic information, as well as the processes of opinion formation, are regarded today as important sources of nonlinear equations describing agents' behavior.

2.1.1. As far as the processing of probabilistic information is concerned, hundreds of experiments have shown that expected utility is no satisfactory representation of human behavior.

In the same way as the *Saint Petersburg Paradox* taught us, in the eighteenth century, that Man evaluates financial rewards with a different scale than that of monetary gains, i.e. with a "psychological" index, in Allaisian words, the *Allais' Paradox* taught us that human beings, when they feel very close to certainty, are powerfully attracted to a completely certain situation. In this vicinity, a one percent change in the probability of some event may mean much more than a corresponding change in the same event when its probability is far from unity. This means simply that Man's behavior under risk is not linear in the probabilities, which is in direct contradiction to expected utility.

Beyond this *"certainty effect"*, other well-accepted characteristics today of Man's behavior under risk have been discovered: the "common consequence" effect (sometimes regarded as a generalization of the certainty effect), the *"Karmarkar Paradox"* or *"utility evaluation effect", the "probability distorsion effect"* [Munier und Abdellaoui, 1991], the *"lottery-effect"* or *"small probabilities overvaluation effect"* etc... [see, e.g., Munier, 1989]. All these results can be synthesized by the proposition that Man's attitude towards risk is not an unambiguous characteristic of the individual and his wealth. Every *risk-structure* is to be related, for a given individual with a given wealth, to a particular degree of risk aversion or of risk proneness. Clearly, such a general result leads to considering expected utility as an erroneous representation of Man's behavior. And it has already been stressed that experimentally observable behavior can be modelled by *nonlinear* equations. When one thinks about multi-period decisions entailing lagged consequences (as, for example, in environmental policy), this result is still reinforced [Machina, 1984, Munier, 1987, Marshall,1989].

2.1.2. There is another important cause for such a nonlinearity. This cause refers to *opinion formation.* Modern economics deals with expectations. However, the fundamental process of opinion formation is more closely linked to memory than to anticipating or forecasting because economic systems are different from physical ones in two respects:

- First, they are full of noise, partly intentionally (mixed strategies in Game theory belong in this category), partly due to existing difficulties in insuring communication in society.
- Second, the observer belongs to the very system he is modeling, with either the power to influence this system or to change the representation

of this sytem, if not discretionarily, at least with many degrees of freedom.

The Bayesian view, which holds that the learning process happens from an "a priori" probability distribution and then from revisions of it through Bayes' Theorem, is thus left with little meaning in economics:

- First, the application of Bayes' Theorem is extremely sensitive to mistakes of the first as well as of the second order as a consequence of the many noises in the system [G. von Fürstenberg, 1990]
- Second, the use of an "a priori" distribution has little meaning from the point of view of the observer, given his position with respect to the system under consideration.

The alternative to the use of Bayes' Theorem leads to considering information as derived from memorizing and from making sense of a careful examination of the environment through a rational deliberative evaluation process. Such processes are usually represented by nonlinear equations in descriptive dynamic systems and are quite often sources for positive feedback.

2.1.3. Finally, physical, financial and other institutional constraints to the economic agent or to society are sources of similarly nonlinear equations. M. Woodford [Woodford, 1986] provides us with convincing examples within the framework of the borrowing and credit mechanism. Examples in the framework of agricultural economics can easily be found and represented through simple models, using even linear objective functions [Munier et Pondaven, 1991, Berger, 1992]. To sum up:

- Human psychology under risk, particularly in cases where decision-making can partly alter the risk structure
- the process of Man's opinion formation
- but also different institutional constraints to human behavior in society

appear to support the idea that the dynamics of decision-making should result in nonlinear equations linking the state variables of a given economic system.

2.2. Endogenous deterministic Uncertainty

These characteristic conditions of different economic activities were hardly mentioned in economic theory until the beginning of the 80's. Today, economists know, however, that such conditions may be sufficient to let dynamic systems exhibit a *deterministic chaos*. Then, economic variables may appear as quasi-randomly fluctuating (Figure 1).

When, in addition, the *dimension of the systems* is large enough, there is no statistical possibility of differentiating the fluctuations of these variables from a "true" random fluctuation process [W. Brock, 1986]. Here, the distinction between complex and complicated takes its full meaning: *Under complexity, any complete understanding of the system is impossible*. This is a major reason why *the determination of numerical probabilities can also turn to be impossible* [Munier, 1986]. What economists characterize as "market efficiency" draws attention to the fact that the dimension of deterministic dynamic systems in the economy is not small. Indeed, if these dimensions were small, the dynamics of the corresponding markets could be deterministically explained and agents informed if they could realize excess profits with certainty. This would destroy one of the strongest justifications of the market economy and would be a direct contradiction to "market efficiency". According to a recent story, a large bank spent $120 million Dollars to "break the code" of the deterministic dynamics of the behavior of foreign exchange dealers. Only later did the management realize that, on epistemological grounds, this could not be done...

Uncertainty appears then as a direct consequence of complexity and behaviors which can be modelled with nonlinear equations.

How, then, can human beings manage in the economy, in the middle of such a complex uncertainty, with the help of bounded rationality? I would like to show, in a third and last part of this talk, that bounded rationality itself provides an unexpected answer to this question.

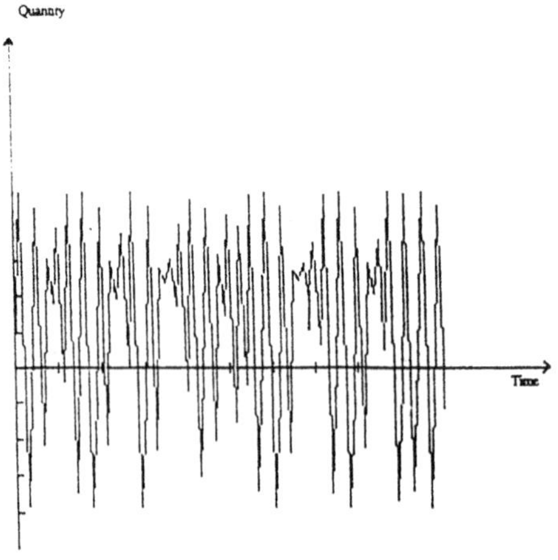

Figure 1. Evolution of the state of a simple one variable linear system endowed with an intertemporal linear constraint

3. BOUNDED RATIONALITY AND THE RETRIEVAL OF HUMAN CAPACITY TO MANAGE ECONOMIC ORGANIZATIONS

What is interesting for us in these complex dynamic models is the *nature of uncertainty*. What we have just seen is that uncertainty in the markets does not always have a purely natural source. When we characterize a decision under uncertainty in terms of a "game against Nature", we follow the old medieval feeling that the ununderstandable - the complexity - is being controlled from the heavens and remains *therefore completely independent from our decisions*. As Albert Einstein's words, engraved on the fireplace mantel at Princeton, illustrate: *"Raffiniert ist der Herrgott, boshaft aber ist er nicht"* (The Lord is sophisticated, but not malicious). This may well be, but uncertainty can nevertheless have a different representation than a purely "natural" one. When, for instance, we deal with price fluctuations on an agricultural market, we too often think in terms of a purely "natural" phenomenon (rain, climate, etc ...): yet, who is not aware that uncertainty in world prices emerges more from the behavior of the other economic agents on the market than from any natural cause!... [Munier et Pondaven, 1991].

I am not referring here necessarily to the so-called "strategic" uncertainty exemplified by the Theory of Games, under which every actor *perceives* that his behavior influences the payments in the game. In the cases mentioned above, the behavior of every agent on the market represents an *unperceived* influence on the *emergence of uncertainty as a product of complexity*. Yet, every agent contributes to this emergence of uncertainty, exactly as, in the case of the price mechanism within the framework of pure competition on a perfect market, every agent does not *intend* to send the price to such or such level, but finally contributes to the equilibrium level reached by the price. This type of uncertainty will be designated from now on by the expression *complex uncertainty*. This phenomenon clearly belongs to the category of *self-organizing systems*.

This result is not neutral with regard to the modelling of rationality. I will try to contribute something here to this problem, to the extent that I will try to provide an answer to the following question: can bounded rationality provide some help, from a prescriptive point of view, to decision makers and in particular managements of organisations, facing complex uncertainty?

3.1. Decisions in Organizations under Complex Uncertainty

Under complex uncertainty, the question is not, in contrast to the already mentioned view of Savage [Savage, 1954], to maximize some expected utility, simply because we "just don't know", in Keynes' words, how to figure out unambiguous numerical values to be attributed to the probabilities

[Munier, 1986]. Dealing with uncertainty must then develop along entirely different lines. Instead of *selecting* the "optimal strategy" as the one bringing the "largest average return in utility", it now makes sense to *look for* - sometimes even to *construct* - a strategy which *preserves a satisfactory flexibility*, given that all sorts of possible events can occur: the organization has to be prepared to take every new course which might be appropriate to meet the dangers.

This can only be obtained, however, through a control of the variety of the organization (the number of potential kinds of behavior of the organization's members) and of the structural flexibility of it (the number of potential changes in the organizational structure). As a general rule, such a control will typically be obtained by influencing the following dimensions:

- the connections within the different functions of the organization
- the connections of the production lines
- the internalization of the organization goals by the available personnel in the organization
- the hiring policy of the organization
- the setting of the strategic decision system of the organization, etc...

The Managers of organizations should, however, take such steps without losing sight of the main objectives of the organization, which should retain a minimal probability of realization.

Descriptively, such behavior can be represented through a maximin rule. Prescriptively, the Simonian I-D-S canonical model will be crucial, particularly in view of the Intelligence and Design phases.

The last observation not only pertains to an organization like a business firm, but also to the entire society.

3.2. Bounded Rationality and the Development of Society under Complex Uncertainty

If one holds, as has just been discussed, that the reaction speed of an organization should be as high as possible, part of the means of the organization will not be one hundred percent efficiently used. For example, the employment policy of a firm may not be oriented toward the most specialized skills, so that everyone on the payroll can shift to different tasks on the spot. In that case, however, to renounce hiring the highest degree of specialized skills may lead to less than one hundred percent efficient use (in the usual meaning of the word, i.e. productivity maximization) of some of the means available to the firm (either human or financial). But with a longer time horizon, the retained flexibility may be the best way to help the organization deal with complex uncertainty: organizational structures will undergo adaptive evolution to the different situations encountered...

What different contributions in the field of self-organization systems have shown [Prigogine und Stengers, 1988, Lesourne, 1991], is that one should expect a satisfactory adaptation of the production system of an economy under these conditions, if (figure 2):

- *First:* Adaptive behavior is actively pursued by the economic agents within the organization
- *Second*: The environment displays complexity
- *Third*: Means can be disposed of in order to fulfill new tasks. This last condition corresponds to the concept of "organizational slack" in the business management literature (used there without relationship to the concept of complexity) and to the bounded rationality in the economic literature (Lesourne, 1991).

These three conditions belong to what could be regarded as a "theorem" of self-organization. The consequently emerging structure of the economy can, in turn, generate under certain conditions complex uncertainty, as the theories of "endogenous" economic fluctuations suggest [Grandmont, 1988].

If individual agents and business firms had, by contrast, used the rule of behavior of "strong" (or "instrumental") rationality, this *adaptation of the economy would not have taken place,* in particular because no means would have been left unused. The uncertainty would presumably have remained at the same level and society would thus not have reached the same level of welfare!

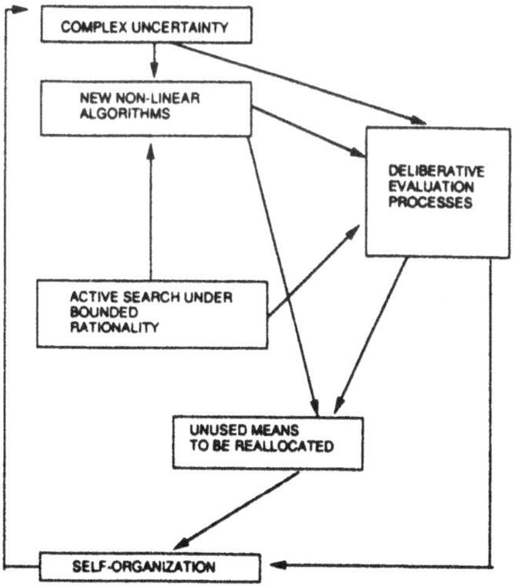

Figure 2. Bounded Rationality and Self-Organization

CONCLUDING REMARKS

In this last section, I would like simply to emphasize why the models, which have been only generally dealt with above, appear more relevant for today's world than those of the conventional static pure economic theory.

The rationality of most decisions in the economy belongs more to the *bounded* type discussed above and has more to do with *dynamic* aspects than the traditional optimization scheme in static or quasi-static models. This latter point constitutes, as a matter of fact, one of the very few differences in basic opinions about economic theory (including game theory) which I had with Oskar Morgenstern - my former professor.

The view I have presented here is perhaps an unusual representation of the economy. It can actually lead to "worsely defined" problems than is the case with the conventional view of uncertainty, but it has, in my eyes, the not negligible advantage of visibly meaning more to businessmen and to individual agents in a market exhibiting complex uncertainty than the conventionally accepted model of strong rationality. Insisting on keeping the latter model would mean that the firm, and more generally speaking Man, would face increasingly complicated computations everyday, which would lead to more disputably "optimal" strategies, with the illusory hope (see above) of overcoming complex uncertainty in the economic world.

In contrast to the traditional view of uncertainty, Man appears here as the *main producer of uncertainty in the economy.* At the same time, Man tries to control this process through a permanent and continuous *adaptation* of organizations (firms and relationships of the individuals to their environment). Ironically enough, choosing an optimal strategy under unchanged organizational structure (or unchanged relationships of the individual to its environment) appears, at this point, as only a second-best possible solution. Establishing this very point, i.e. that Man's bounded rationality *achieves unexpectedly more in society under complex uncertainty* than the "strong" rationality scheme of traditional economics has been my main objective in this talk.

Prof. Dr. Bertrand Munier
Ecole Normale Supérieure de cachan
Departement d'économie
61, Avenue du Président Wilson
94235 Cachan Cédex, France

REFERENCES

Allais, M. [1953], "Le Comportement de l'Homme rationnel devant le Risque, Critique des Postulats et Axiomes de l'Ecole Américaine", *Econometrica 21*, no. 4, pp. 47-73.

Allais, M. [1979], "The So-Called Allais Paradox and Rational Decisions Under Uncertainty", in: Allais, M. and O. Hagen, eds., *Expected Utility Hypotheses and the Allais Paradox*, Dordrecht/Boston: Reidel.

Berger, E. [1992], *Etude d'un Marché agricole non soutenu*, Note de Recherche GRID no. 92-02, E.N.S. de Cachan, Mimeo.

Brock, W.A. [1986], "Distinguishing Random and Deterministic Systems", *Journal of Economic Theory 40*, no. 1, pp. 168-195.

Brock, W.A. and W.D. Dechert [1988], "Theorems on Distinguishing Deterministic Systems", in: W. Barnett, E. Berndt and H. White, eds., *Dynamic Econometric Modelling*, Cambridge: Cambridge University Press, pp. 247-265.

Grandmont, J.M. [1988], *Nonlinear Difference Equations, Bifurcations and Chaos: An Introduction*, Note de Recherche CEPREMAP 88-11, Paris.

Keynes, J.M. [1936], *The General Theory of Employment, Interest and Money*, London: Reprinted Harbinger, Harcourt Brace and World 1964.

Knight, F. [1921], *Risk, Uncertainty and Profit*, New York and London: Reprinted University of Chicago Press 1971.

Laux, H. [1982], *Entscheidungstheorie*, Berlin: Springer.

Le Moigne, J.L. [1977], *La Théorie du Système Général, Théorie de la modélisation*, Paris: Presses Universitaires de France.

Lesourne, J. [1991], *Economie de l'Ordre et du Désordre*, Paris: Economica.

Machina, M. [1984], "Temporal Risk and the Nature of Induced Preferences", *Journal of Economic Theory 33*, pp. 199-231.

Marshall, John M. [1989], "Welfare Analysis Under Uncertainty", *Journal of Risk and Uncertainty 2*, no. 4, pp. 385-403.

Mongin, Ph. [1984], "Simon et la Théorie néo-classique de la Rationalité limitée", in: Demailly, S. and J. L. Le Moigne, eds., *Sciences de l'Intelligence, Sciences de l'Artificiel*, Lyon: Presses Universitaires de Lyon.

Mongin, Ph. and B. Walliser [1988], "Infinite Regressions in the Optimizing Theory of Decision", in: Munier, B., ed., *Risk, Decision and Rationality*, Dordrecht/Boston: Reidel, pp. 435-458.

Munier, B. [1986], "Complexité et Décision stratégique dans l'Incertain: Que peut-on retenir de la Théorie?", in: Boiteux, M., Th. de Montbrial and B. Munier, eds., *Marchés, Capital et Incertitude*, Paris: Economica.

Munier, B. [1987], "On the Inadequacy of Expected Utility in Selecting Public Investments", in: Rand, G., ed., *Operational Research 87*, Amsterdam, North-Holland, pp. 843-852.

Munier, B. [1989], "Calcul économique et Révision de la Théorie de la Décision en Avenir risqué", *Revue d'Economie Politique 99*, no. 2, pp. 276-306.

Munier, B. [1989], "Décision", *Encyclopaedia Universalis VI*, pp. 74-80.

Munier, B. [1991], "Market Uncertainty and the Process of Belief Formation", *Journal of Risk and Uncertainty 4*, no. 2, pp. 233-250.

Munier, B. and M. Abdellaoui [1991], "Expected Utility Violations: An Appropriate and Intercultural Experiment", in: Chikàn, A. ed., *Progress in Decision, Utility and Risk Theory*, Dordrecht/Boston: Kluwer Acad. Publ.

Munier, B. and C. Pondaven [1991], *Portée et Limites de la Stabilisation des Prix agricoles*, Rapport au Commissariat Général du Plan, Paris: Mimeo GRID.

Prigogine, I. and I. Stengers [1988], *Entre le Temps et l'Eternité*, Paris: Fayard.

Savage, L J. [1954], *Foundations of Statistics*, New York: Wiley. 2nd edition: Dover Publications, 1972.

Simon H.A. [1955], "A Behavioral Model of Rational Choice", *Quarterly Journal of Economics 69*, no. 1, pp. 99-118.

Simon, H.A. [1976], *Methods and Appraisal in Economics*, Cambridge: Cambridge University Press.

Simon, H.A. [1978], "Rationality as Process and as Product of Thought", *American Economic Review 68*, no. 1, pp. 1-16.

Simon, HA. [1979], "Rational Decision Making in Business Organizations", Nobel Lecture, *Americain Economic Review 69*, pp. 494-513.

Simon, H.A. [1983], "The Role of Expectations in an Adaptive or Behavioristic Model", in: Simon, H.A., *Models of Bounded Rationality*, Cambridge, MA: Harvard University Press, pp. 380-399.

Stigler, G.J. [1961], "The Economics of Information", *Journal of Political Economy 69*, no. 2, pp. 213-225.

Teilhard de Chardin [1959], *Le Phénomène humain*, Paris: Flammarion.

Von Neumann, J. und O. Morgenstem [1947], *Theory of Games and Economic Behavior*, Princeton: Princeton University Press, 2nd edition.

Winter, S. [1975], "Optimization and Evolution in the Theory of the Firm", in: Day, R.H. and T. Groves, eds., *Adaptive Economic Models*, New York: Academic Press, pp. 73-118.

Woodford, M. [1986], "Stationary Sunspot Equilibria in a Finance Constrained Economy", *Journal of Economic Theory 40*, no. 1, pp. 128-137.

SECURITY. ON THE REASONS FOR THE SINKING ACCEPTANCE OF RISK

HERMANN LÜBBE

The subjects "risk" and "security" are currently a great and ever increasing concern of the sciences involved. The number of titles in both specialist and non-specialist literature is growing and demonstrates the actuality of these topics. Universities devote series of interdisciplinary lectures to them. Major research institutes and their study groups are preoccupied with the relevant questions, quite simply for the reason that they are affected. Obviously, this also applies to industries and their federations including the associated professional organizations.

The question as to the reasons for the actuality of the topics "risk" and "security" is becoming insistent. The answer to this question may appear to be trivial. In this actuality, the growing desire for security of people living in modern industrial societies is reflected, and with this desire the population is reacting to the experience of the objectively increasing risks it is exposed to in our industrial society.

Mere mention of the key word "Tchernobyl" seems to suffice to confirm the thesis of the objectively accumulating burden of risks modern industrial man must bear. Yet Tchernobyl is just the prototype of those technical catastrophes, several of which any up-to-date media consumer could enumerate spontaneously - from Bhopal to Seveso and Basel. At the chemical accident in Bhopal, the number of victims was in fact more than ten times higher than in Tchernobyl. And it is also known that the drinking water supply of hundred thousands of residents near the Rhine is ensured, last but not least, by bank filtrates. In view of these facts, how should a chemical accident at the upper reaches of the Rhine not be highly alarming, especially when confronted on television with pictures of fish carcasses dozens of kilometers down the river?

So it seems that in the growing desire for security, the objectively growing risks of life of the population in modern industrial society are reflected. On the other hand, a good indicator of actually given life security might surely be our average life expectancy. Taking this as a measure, people in modern industrial societies would be living more securely than man has ever lived before. Medical know-how, the pharmaceutical products of chemical industry, above all, however, hygiene - itself contingent upon welfare - have on the whole made our lives safer, at least where health is concerned. Due to the achievements of modern technology and organization, we are nowadays also better protected against natural disasters than in earlier times. In which respects then are we now living under higher risk than we used to?

J. Götschl (ed.), Revolutionary Changes in Understanding Man and Society, 243–252.

Be that as it may, quite independently of the way we answer the question whether we are now, in modern industrial society, existing objectively in more security or more risk than in earlier ages - in highly developed industrial societies the acceptance of risk is sinking. The reasons for this are elementarily connected with the facts of modern life. Their effects may be considered irreversible, and pep talks under the motto "dare to risk!" are probably to no avail whatsoever; they must be dealt with such as they are.

Which are the reasons implied? I would like to confine myself to making six of these reasons plausible as to their effects on the desire for security and risk acceptance.

First. As the extent of those conditions of life which are produced by ourselves increases, the willingness to meekly bear the risks of life sinks. This can be exemplified by the consequences for our well-being due to progress in medicine. Advances in prenatal diagnostics are a point in case. The number of expectant mothers who make use of the possibilities of this progress - usually with a view to being able to induce an abortion, should the child in the womb be found to be damaged in a way not susceptible to therapy - is continuously growing. But of course here too, medical diagnostical art remains fallible in principle, so that it can occur, although rarely, that one finds oneself born into this world severely impaired, instead of healthy as predicted.

The number of such cases is growing, and the cases have become correspondingly frequent in which children born with handicaps and represented by their parents and lawyers are sueing the fallible doctor for damages before the civil judge for the injury their own handicapped existence, which has become a fact against all intentions, signifies in contrast with the higher value they attribute to their non-existence. It is comforting to hear that with the exception of a single case in the United States, liability of the doctors implicated by their incorrect diagnosis for compensation of the damage implied by one's own existence ("wrongful birth") has regularly been refused in our courts.

At any rate, in the cases mentioned our civil courts do usually allow claims for damages regarding the special expenditures resulting from having to live a life given birth to with handicaps against all intentions, and these are damages, after all, which could only have been avoided by the timely transfer of the existence in question into non-existence.

One way or the other, the ever growing number of children eagerly awaited or not born at all due to progress in reproductive medicine, is an especially impressive example of the civilization-specific transformation of the conditions and facts of life into results of acts, and it is evident that corresponding to this process, the readiness to accept risks in life must sink - in the above-mentioned case, the self-acceptance of one's own existence now perceived as a risk. Bad consequences of acts are eminently less acceptable than the bad consequences of merely natural processes, and the process of civilization is in fact a process of transformation from the latter into the former. This has an irreversible effect on our behaviour regarding the acceptance of risk, quite independently from the answer to the question whether our life has become either more or less secure within the context of industrial society.

Second. The experience of risk becomes more intense as the intervention of our technically implemented acts gains in importance both naturally and socially, and as our knowledge grows concerning the unintended effects of what we do when acting technically with generally approved intentions. In modern societies we make the experience of increasing dependence on the action of other, remote people. This too has an irreversible effect on our state of mind. It is not only true that bad consequences of acts afflict us more deeply than bad consequences of merely natural processes. It is also true that risks to which we are exposed as a consequence of the acts of others afflict us more deeply than risks we must attribute to our own agency. Accordingly, together with the extent of our real dependence on the acts of other people in modern society, our claim addressed to these same other people for having our security ensured is expanding. Already by participating in modern street traffic, we are all of us aware of this connection. The tendency to blame oneself when causing bodywork damage usually quickly wanes. When others are responsible for the accident, we claim for damages quite independently from the moral and criminal quality pertaining to the damaging action of these persons in question, and the legal system complies with our claim.

Besides, as the natural and social importance of our intervening, technically implemented acts grows, the extent of damage in the case of a catastrophe also tends to grow. But the extent of the damage due to catastrophes now menacing us is not identical with the magnitude of the risks we are taking. This too is well-known to anyone participating in street traffic. Clearly, the extent of damage in accidents such as those with cars potentially includes one's own death, and in fact, every year ten thousands of people die in the traffic in Europe. Nonetheless, every day millions get into their cars with hardly a qualm - not because they audaciously disregard the extent of damage their own death implies, but because, from a practical point of view quite correctly, they consider the probability of being the ones to die in a street accident bearably low. Also, someone unacquainted with the technology of security is thus able to see, from an everyday perspective, that the risks he is exposed to are not identical with the extent of damage due to the catastrophes which might happen. He quite realizes from a practical point of view that it is the potential extent of damage in combination with the probability of its occurring which alone constitutes the risk.

However, the extent of damage always impresses us more than the risks. Media reports demonstrate this. An average of twenty car accident victims per day is not worth reporting. Yet a single little airplane accident with the same number of victims requires headlines - such is the case, once and again. May one - not only with regard to specifically civilized individual, but also to collective life - expect us in future, when assessing the security of our lives, to learn to orientate ourselves to risks, instead of to impressive extents of damage? For the time being, it is probably more realistic to assume that the so-called acceptance of risk especially sinks with the growing extent of damage due to catastrophes which potentially menace us.

Third. The experience of insecurity increases correlatively to the magnitude of the loss of experience specific to civilization. What is meant by "the loss of experience"? This can be elucidated by a look back at the prehistory of industrial society. Before the process of industrialization in the modern sense started with the installation of the first steam engine two hundred years ago of British manufacture, in European societies, between two thirds and three quarters of the people depended on agriculture for a living. It would be quite inappropriate to romanticize this fact. Merely looking at the average life expectancy in those times forbids this, for it only amounted to half of ours. Other aspects of life in agrarian societies can, however, certainly be considered advantages from our present point of view, for example the fact that in those times a great majority of the population had a relationship with the real conditions of physical and social existence marked by rich personal experience and concomitant stability. At the same time, the degree of economical and social autarky was considerable - as evidenced by the fact that going to market was seldom necessary. Emphatically speaking, people knew life. If, in contrast, we ask ourselves what we now - and this means every one of us - still actually know through experience of life about the real conditions of our physical and social existence, it becomes clear that there never has been a civilized society which understood its conditions of life less than our own.

It follows from this that now, more than ever before, we have to rely on trust - trust in the simple sense of trusting the soundness of what the expert respectively beside us does. One can become fully aware of our dependence on trust in precisely the above-mentioned sense by visualizing the abundance of acts of trust to be found within a single day and without which we wouldn't be able to live - from trust in the scientifically based art of the dentist whom we visit in the morning before work, to trust in the good working order of the pocket calculator whose results an engineer constructing bridges relies upon for his static calculations.

In short, in complex and highly mobile societies we compensate for the evaporating extent of our own common judgement with expert knowledge we can trust. In the political system this expert knowledge was long ago institutionalized, and the social scientists for their part long ago made surveys of this. There is no scientist of importance above the average who is not in one such expert council or more. Thus, expert knowledge compensates for common sense's waning extent of judgement - as glasses compensate for the waning extent of sight in myopia.

As far as compensations work well, there is no objection implied in saying that they are only compensations. Things always get precarious when the expert counsel we ought to be able to trust in complex and dynamic societies loses trustworthiness. Precisely this, however, is always the case when the experts themselves are unable to agree, even showing signs of mutual bitterness. The number of cases in which this happens is growing. Particularly in great public hearings such as those which governments and legislative corporations hold in ever larger numbers, this is the case, all the more frequently as the technical and organizational complexity of the decisions for whose political preparation the expert counsel is required increases. The trust referred to above, which is becoming more necessary as a socially cementing agent for the reasons mentioned, then crumbles and loses its binding power.

The obvious political form of reaction to experience concerning the waning extent of common sense together with sinking trust in expert judgement, is to refrain from judgement. "How do politics react to the acceleration of contemporary history?" - such is the subtitle of the impressive study by the political scientists Gruner and Hertig from Berne, who render the fact plausible that moratoriums on judgement must increase in the conduct of the voter, if in view of the complexity of decisions to be taken on the one hand and lacking concurrence of the experts on the other he should become overtaxed. This is the structure of a situation in which using slogans such as these reported from the United States in voting campaigns becomes plausible: "Confused? Many are. Play safe! When in doubt, vote No!" As one can see, this No is not the No of well-founded rejection, but the No of abstained judgement under pressure of the experienced overtaxation of one's own power of judgement - the moratorium-No, as we might call it. Generally, in modern, highly complex societies the trend to this No seems to be increasing. Nota bene: this No is not a case of irrational reaction, but a rational reaction one must count on in all fairness. Insofar as this is correct, it follows that modern societies can come to terms with undoubtedly high degrees of complexity and change-related dynamics reaching far into their political systems, but not to an unlimited degree. Beyond certain limits, trust in the reliability of what happens in our black-box civilization so dependent on the performance of experts is overtaxed - with the observed destructive consequences for our attitude to acceptance.

Fourth. The experience of insecurity becomes more intense at the rate with which the range of information accessible to us through the media becomes greater than the range of action within which we can take dispositions. That extending the range of what we get information on exacerbates the experience of insecurity only seems a paradox at first glance. Upon second sight, the potency of information as a medium for the experience of impotency specific to civilization is recognizable. The term "world society" is well-known by now. In many respects of life, it would be an exaggeration to say that we are already now in a world society. But in at least one respect, dependent on technical developments, what may be called a "world society" does actually exist already, namely as the information society. To begin with, this merely means that the globe has by now practically become an integrated system of telecommunications technology. In the shortest time, the public throughout the world is nowadays informed on events and occurrences of importance, and populations lacking receiving appliances and not yet within reach as addressees of medially propagated information, and thereby still living outside the world public, are dwindling. It is true that by far the greater part of so-called world news is without any practical relevance whatsoever for us, that it doesn't increase our competence, doesn't require any reaction, and so means no more than entertainment.

But to a small, very special class of information exactly this does not apply. In its spatial, ultimately global extension, the process of industrialization implies the increase of wide-ranging actual interdependence - from the trade routes over which we are supplied with raw materials and food and energy sources, to the communications systems upon whose functioning in this era of absolute weapons world peace, ensured by mutual deterrence, is dependent, to the cooperative capabilities of those multinational committees without whose counselling and decision-making work neither the safety of world-wide airplane travel could be ensured nor ecological catastrophes across the borders be avoided. In this world of actually expanding mutual dependence, the confrontation of populations in distant areas lying even beyond Turkey, is anything but a reason to realize the contrasting comfort of one's own situation. This was the effect after all that even Goethe used to characterize the petit bourgeois pleasure of reading newspapers. In contrast, Golf and Channel wars today are triggers of acute as well as chronic experiences of insecurity. Politically caused energy shortage threatens, as do energy price increases, and only superpowers dispose of relevant political means to influence the menacing events. The citizens of small as well as medium-sized states and the governments of these states however experience their own impotence faced with processes and events which after all do potentially and very actually affect them.

Evidently, the synchronous medial presence of information on all world events with global importance, together with the actual dependence on these events which can hardly be influenced, is a combination pregnant with the experience of impotency.

Fifth. The experience of insecurity becomes more intense with decreasing social control. The effects of decreasing social control have by now become part of our daily life. Older tourists remember that three decades ago the doors of rural churches were still regularly open. Now you will find them shut in even the remotest of valleys. If they weren't closed, mediocre peasant-art putti would also not be in their place at the altar for long, let alone rustic madonnas. Reverential piety hardly hinders the illegal antiques trade any more, and under cover of specially constituted tourist crowds, one can quite easily get away with stealing a little stone from a Byzantine or even Roman mosaic as a souvenir.

Feelings of being menaced arise when in rooms of first class hotels around the world, strict instructions must be observed not to open the door upon knocking before having looked through the spyhole. In many city districts house owners long ago took analogous precautions for granted. There are cosmopolitan cities in which it is unwise to take an evening walk alone. The police advises people to carry neither a lot of money nor absolutely none with them in the subway, but a little money, so that in case of being assaulted damage on the one hand is limited and on the other, muggers aren't provoked to aggression because of frustration.

One could continue for a long time in the description of such changes in the way we lead our lives. What is the reason for these changes? The sociological key word "social control", and this means decreasing social control, is an answer. Explaining the issue thus named is a complicated matter. What is meant becomes clear when one considers a relevant aspect of social modernity. In the development of modern, differentiated, and complex societies, the relative importance of our direct mutual social dependence decreases, while complementarily, our indirect social relations of dependence expand. The range of our social relations grows simultaneously with the functional differentiation of these relations. Hand in hand, our subjectivity in the emphatic sense is withdrawn from these relations. With an increment of privacy and intimacy it recedes into small and smallest groups. In other words, with the expansion of the public spheres in which we move socially, we are increasingly provided with chances to do this under cover of anonymity. This is the way in which, more than one hundred years ago, the phenomenon "city" as a social phenomenon was analysed.

Under the protection of large modern societies, the number of possibilities of making oneself impossible decreases. This implies that conformities in behaviour which are guaranteed by social control dissolve, the formation of subcultures is encouraged, individuals develop their own personalities more fully than ever before, and also for activities in the sphere of lesser criminality which make the experience of insecurity an everyday matter, there are unprecedented chances of development.

From the standpoint of cultural history, this may be exemplified by the complementary process of technical developments in locks and other safeguarding installations. To put it clearly, a society rich enough to make owning a second house possible to an unprecedented degree, must not wonder if the house which is not inhabited at the moment is a favoured object of organised burglary, and the above-mentioned items offered by the expanding technological security industry serve to compensate for this. These observations can also be transferred to the political sphere of life. With the modernity of modern societies, the amount of convictions considered binding for everyone is diminishing. Complementarily, the pluralism of opinions is growing, last but not least where conditions of life are concerned which, because of their degree of scientific and technological complexity, are increasingly difficult to comprise within the range of competence of our common-sense judgement. The spectrum of public opinion in matters of technological politics is not last to be accordingly differentiated, and opinions also diverge as to the political activities justified by one's own convictions. And moralists, whose number tends to grow with the inscrutability of affairs, in extreme cases do not doubt their higher right to violence. A consequence is that, in constantly growing public contexts, we must hope to find security by installing technical security systems. Before we board the plane on a business trip, we must submit to electronic frisking. Baggage found without owners results in the bomb experts being called. Large nuclear power plants or chemical works are armed like fortresses - with walls topped by barbed wire, gates fortified with steel to provide against the attacks of armoured terrorist vehicles, with communications trenches for patrols with and without dogs, electronically surveyed and illuminated by floodlights at night. In short, never before in the history of civilization have we been surrounded daily by such security measures. It would be unrealistic to assume that we felt all the safer for this reason. The contrary is obviously the case. The number of citizens considered by other citizens to be those against whom we must be protected is growing.

Sixth. The desire for security grows with the rising level of technical and social security. Once again, it only appears a paradox that our sensibility concerning restrictions of our security increases with the degree of the security level attained. The history of medical culture may serve to exemplify what happens here. The famous and notorious definition of health by the World Health Organisation, according to which health is a state of unrestricted physical, psychological, and social well-being, would have been considered presumptuous, even blasphemous until deep into the 19th century. Clearly, the historically unprecedented achievements of modern medicine, in combination of course with the historically unprecedented economical efficiency of modern industrial society, are what has raised the level of our demands in matters of health to the current heights. In approximation to the conceivable optimum, the remaining gaps quickly gain in intolerability. This process may be open to closer psychological explanation. In everyday practice we know this by looking at the course our state of mind takes with each drive home: the closer we are to our aim, the more impatient we get.

But it would be a misunderstanding critical of civilization to assume that our growing claims to security mirror an increasing decadence in our capacity of self-determination. In this respect, the very opposite is the case. Thanks to welfare and the life-span relieved of necessity at our disposal, the extent of opportunities for leading life in a self-determined way has never been greater than today, and these opportunities are taken advantage of. This is reflected in the so-called change of values, and "self-realisation" as a life-orientating value does not represent decadent egotism, but rather an objective necessity in view of the unprecedented expansion of the sphere within which we can dispose freely, and within which nothing at all would happen if it didn't happen in a self-determined way. Of course everyone knows that it is not the social autarky of individuals or small groups that the liberties of modern life, which are appreciated and generally put to good use, are due to, but the social security modern society alone can guarantee with its political institutions, and for this reason the demands on the system of our social security are growing - not in spite of the claims to liberty of modern citizens, but because of them.

Prof. Dr. Hermann Lübbe
Philosophisches Seminar
Universität Zürich
Rämistraße 71, 8006 Zürich, Suisse

REFERENCES

Beck, Ulrich, *Risikogesellschaft*, Auf dem Weg in eine andere Moderne, Frankfurt am Main 1986.
Crescenzo-d'Auriac, Marie-Béatrix, *Les risques catastrophiques*, Evénement naturel, politique et technologique, Préface de Jacques Lallement, président de la Fédération francaise des sociétés d'assurance, Paris 1988.
Deucher, Wolfgang, *Die Haftung des Arztes für die unerwünschte Geburt eines Kindes ("wrongful birth"*, Eine rechtsvergleichende Darstellung des amerikanischen und deutschen Rechts, Frankfurt a.M., Berlin, New York, Nancy 1984.
Halpérin, Jean, *Les assurances en Suisse et dans le monde*, Leur rôle dans l'évolution économique et sociale, Neuchâtel 1946.
Harrington, Scott E., "A Retrospective on the Liability Insurance Crisis", in: *CPCU Journal*, 1990 pp. 17-28.

Kaufmann, Franz-Xaver, *Sicherheit als soziologisches und sozialpolitisches Problem*, Untersuchungen zu einer Wertidee hochdifferenzierter Gesellschaften, Stuttgart[2] 1973.
Knörr, Karl, "Pränatale Diagnostik - Klinik und Folgerungen", in: Odo Marquard, Hansjürgen Staudinger (Hrsg.), *Anfang und Ende des menschlichen Lebens*, Medizinethische Probleme, München, Paderborn 1987, pp. 24-36.
Krejci, Heinz, "Schäden durch internationale Katastrophen im Lichte des Schadenersatz- und Versicherungsrechtes. Österreichischer Landesbericht anläßlich des 8. Weltkongresses der Internationalen Vereinigung für Versicherungsrecht 1990 in Kopenhagen", in: *Die Versicherungsrundschau. Zeitschrift der Österreichischen Gesellschaft für Versicherungsfachwissen 45*, 1990, pp. 115.
Layton-Cook, Nanci, "Medial Malpractice Insurance", in: *Professional Liability Today* 1990, pp. 2-3.
Lindner, Roland, "Technik und Gesellschaft IV. Risikoeinschätzung und Akzeptanz neuer Technologien. Kommission der Europäischen Gemeinschaften", Bericht *EUR 9179 DE* 1984, pp. 6ff.: "Gibt es einen allgemeinverständlichen Risikobegriff?".
Lübbe, Hermann, *Der Lebenssinn der Industriegesellschaft*, Über die moralische Verfassung der wissenschaftlich-technischen Zivilisation, Berlin, Heidelberg 1990, pp. 56ff.
Lübbe, Hermann, *Politischer Moralismus*, Der Triumph der Gesinnung über die Urteilskraft, Berlin[2] 1989.
Scharlau, Martin, "Die Entstehung neuer Versicherungszweige", *Veröffentlichungen des Deutschen Vereins für Versicherungs-Wissenschaft*, Hrsg. von Alfred Manes, Heft XLIII (ausgegeben Januar 1929), Berlin.
Schneider, Th, "Grundgedanken und Methodik moderner Sicherheitsplanung", in: *INTERPRAEVENT*, Band 1., Wels (OÖ) 1980, pp. 49-69, p. 50.
Swiss Reinsurance Company: Publications, Zürich 1993.
Taschner berichtet über europarechtliche Angleichungstendenzen: *Produkthaftung*, Richtlinie des Rates vom 25. Juli 1985 zur Angleichung der Rechts- und Verwaltungsvorschriften der Mitgliedstaaten über die Haftung für fehlerhafte Produkte (85/374/EWG). Erläutert von Hans-Claudius Taschner, M.C.J., Abteilungsleiter bei der Kommission der Europäischen Gemeinschaften, Brüssel, München 1986.
Zapf, Wolfgang, Sigrid Breuer, Jürgen Hampel, Peter Krause, Hans-Michael Mohr, Erich Wiegand, *Individualisierung und Sicherheit*, Untersuchungen zur Lebensqualität in der Bundesrepublik Deutschland. München 1987, p. 3: "Einem gestiegenen Individualisierungsdruck korrespondiert ein steigendes Sicherheitsbedürfnis".

PART V

AESTHETICAL AND ETHICAL PATTERNS

NATURE AS A WORK OF ART

PAUL FEYERABEND†

1. INTRODUCTION

The topic of the conference is "Towards a Theory of Man and Society". My own view is that neither humans, nor society nor even nature (whatever that is) can be captured by theories; that the statements, equations, formulae which apparently do the job are successful for historical reasons, not because they reflect an objective world; and that whatever universality we find comes from the actions of scientists, politicians, social reformers who, having immersed themselves in various problem situations and using their experience, cunning and other forms of 'tacit knowledge' now happen to treat different cases as if they were the same: universal standards of scientific knowledge and the universality implied by scientific theories (of nature, society, humans) are contingent features of scientific (political, social) practice, they are not transhistorical agencies.

Among philosophers this view is customarily called relativism, or nominalism, or historicism. It has been around for a long time and it was often connected with important social and political trends. Protagoras' statements that man is the measure of all things (Plato, Theaetetus 152a2f) displeased not only specialists; it also clashed with the idea, central to conservative thought, that the laws of a city were of divine origin and should not be changed. Laws, according to Protagoras, are necessary for order: whoever breaks them not just once, but repeatedly must be "put to death as a plague to the city" (Plato, Protagoras 320ff). Still, they are not sacrosanct. New conditions and new wishes require new laws and new forms of knowledge. This 'relativism' has little in common with modern movements of the same name. It employs experts (physicians, philosophers, politicians etc.) and explains how an appeal to them can be reconciled with the man-measure doctrine (Plato, Theaetetus 166dlff). Plato wanted greater stability. A good society, he thought, must agree with more than the wishes of its citizens; it must conform to the features of an objectively existing real world and, ultimately, of the entire universe (Timaeus). Plato and Protagoras also dabbled in practical politics, the one in Sicily, the other, according to some reports, by devising a constitution for Thurii, a Panhellenic settlement in Southern Italy. Thus an ancient debate between 'realism' and 'relativism' had major political implications.

J. Götschl (ed.), Revolutionary Changes in Understanding Man and Society, 255–267.

2. PARMENIDES

An interesting and truly fascinating feature of early Greek thought is that this relation between philosophy, cosmology and politics could be reversed; political judgements could be used to modify the study of nature and Being. The Ionian philosophers had postulated a unity behind diversity and had tried to reduce observable events to it. Thales, according to Aristotle, identified the unity with water. We can assume that he had arguments for his choice, for example, that water can turn into solid, airy and fluid matter and that it is necessary for life. Anaximander replaced water by an indefinite substance which could appear in different forms and which he called apeiron, while Anaximenes opted for air, again using plausible arguments. Parmenides pushed the tradition to the extreme and clearly showed its implications. According to Parmenides the entity sought is neither water, nor apeiron, nor air; it is Being for Being, uniting all particular substances. What can we say about Being? That it is (estin)[1] and that not-Being is not. What happens on the basic level? Nothing: the only possible change of Being is into not-Being, not-Being is not, hence there is no change. How is the basic level structured? It is full, continuous and without subdivisions. Why? Because any subdivison would be between Being and something else, the only something else on the basic level is not-Being, not-Being does not exist, hence there are no subdivisions.[2] But is it not true that we assume and experience change and difference? Yes, we do. Which shows, according to Parmenides, that neither tradition nor experience (e'thos poly'peiron - 87, 3) are reliable guides to reality. They are chimeras. So far for Parmenides' argument.

Those who feel inclined to ridicule the argument should consider that large parts of modern science contain a howdlerised version of its result. Classical physics distinguishes between a 'real' world without colors, smells, sounds etc. and an everyday world in which such qualities play an important role. The real world does change, but the change is not very profound; it can be reversed and it merely transports initial conditions from one moment to another. Einsteinian reality is even more rigid. All events are laid out in advance. This world "simply is, it does not happen. Only to the gaze of my consciousness, crawling upward along the lifeline of my body, does a section of this world come to life as a fleeting image in space which continuously changes in time." "For us", wrote Einstein, "who are convinced physicists the distinction between past, present and future has no other meaning than that of an illusion, though a tenacious one".[3]

Irreversibility was often ascribed to an interaction between nature and the observer, not to nature itself. Molecular biologists despise subjects such as botany and zoology where the observation of qualities plays a major role. And so on. None of these scientists can offer arguments which are as simple, clear and compelling as those of Parmenides and nobody has explained how deceptive appearances can inform us about a real world that excludes them. Compared with Parmenides, the Parmenidean scientists and their philosophical supporters seem like children. But their ideas, though watered down, still show the effects of the vision[4] of the Great Unifier.

3. ARISTOTLE'S CRITICISM

Aristotle criticized Parmenides in two ways. He tried to show the mistakes in Parmenides' reasoning and he pointed out that change is important in human life. 'Even if there existed a Good that is one and can be predicated generally, or that exists separately and in and for itself' he wrote about similar ideas in Plato (EN 1096b33ff - my emphasis)

> it would be clear that such a Good can neither be produced nor acquired by human beings. However it is just such a Good we are looking for ... one cannot see what use a weaver or carpenter will have for his own profession from knowing the Good in itself or how somebody will become a better physician or a better general once 'he has had a look at the idea of the Good' [apparently an ironical quotation of a formula much used in the Platonic school] It seems that the physician does not try to find health in itself, but the health of human beings or perhaps even the health of an individual person. For he heals the individual.

This is a most interesting passage. Assume, says Aristotle, that there are unchanging entities. Assume furthermore that our lives, though stable to some extent, don't show a trace of them. Then we can either try to adapt to the entities, or continue on a way that has the strength of experience and tradition behind it. The first alternative is part of religious cults. Taking particular experiences as their starting point and utilizing the inherent ambivalence of phenomena (a point I am not going to discuss in the present essay), the members of such cults reorient their thinking and their lives. According to Aristotle the citizens of Athens already made their choice. They do not merely accept change, they maintain it by their actions. "For unlike any other nation" said Pericles in his funeral oration, as reconstructed by Thucydides (The Peloponnesian War, II, 40) "we do not regard him who takes no part in [civic] duties as unambitious [or as an expert in Being] but as useless." Using the word 'real' to refer to what is basic for an individual, a group, or a nation we can say that change, for the Athenians, was very real indeed; or, generalising: real is what plays a central role in the kind of life we identify with. I shall call this generalization Aristotle's principle.

Aristotle's principle is not a philosophical principle in the sense that it disregards habit and tradition and descends to a more hidden and more profound kind of reality. It "leaves everything as it is" (Wittgenstein, Philosophical investigation, 124). It stays with habit, commonsense and tradition, describes how they are divided into what is important not only to the minds of people but for preserving the form of life that surrounds them and what is more peripheral and idiosyncratic; and it applies the word 'real' to things, events, processes of the first kind. Compared with this procedure, a philosophy, or a religious view that declares a special invention, an extrapolation or isolated phenomena (such as the crucification of Christ) to be manifestations of a deeper and largely hidden essence sounds rather Utopian. In the extreme case - no connection whatsoever with what is known - the reality proposed is indeed beyond understanding. Mysticism and negative theology are close to the extreme case but not identical with it. The intended reality occasionally shows itself and can be hinted at by symbols, images, analogies. The early Greek philosophers were in an even better position. When Xenophanes mocked the Homeric gods because they had human and tribal idiosyncracies the difference between local deities had already been washed out to such an extent (by travelling, international communication and other equalizing agencies) that the mockery started looking like an argument.[5]

Similarly Parmenides' notion of Being (it has not-Being as its only alternative) sounded plausible in a period when social life became increasingly abstract: money replaced gift-giving and the exchange of goods, local gods merged, gained in power but lost in concreteness and humanity, abstract laws, not personal relations defined the role of a citizen in a democracy, wars were fought by professional soldiers and so on. There was social strength in Parmenides' considerations or else they would have been incomprehensible. But, Aristotle seems to say, this strength is overruled by the strength of change and diversity.

4. APPLICATION TO QUANTUM MECHANICS

An interesting modern example of the application of Aristotle's principle and of the fact that it does not always lead to unique results is quantum mechanics. Taken literally, quantum mechanics has a consequence which is very similar to that of Parmenides': there are no well-defined objects and no distinguishable observations which can be stored. Movements that find salvation in overcoming change and difference welcome this feature and want to preserve it. Bohr, being more a man of the world, rejected the consequence. The observer, he said, must be separated from his surroundings and described in commonsense or, if precision is needed, in classical terms. Like Aristotle he made a practical decision and interpreted quantum mechanics accordingly.

In Bohr's interpretation, human beings function as 'detached observers'. But we don't experience ourselves, our actions, and our relation to the world in this detached way. Bohr himself often emphazised that we are both actors and observers on the stage of life. Wolfgang Pauli tried to reconcile physics with this more 'subjective' view of human beings and to reinstate cultural elements that had been eliminated by the advance of the sciences. He therefore preferred a more literal reading of quantum mechanics. "According to my point of view" he wrote (letter to Bohr of Feb. 15, 1955) "the degree of [the] 'detachment' [of the observer] is gradually lessened in our theoretical explanation of nature and I am expecting further steps in this direction".[6] We see here very clearly how the weight ascribed to certain phenomena and the wish to give them a larger role within an important enterprise leads to attempts to make room for them, to regard them as "real" and how such attempts in turn influence the interpretation of scientific theories: it makes no sense to speak of "the objectively correct interpretation" of a particular theory. Let us now see what happens when the procedure is applied to cultures or historical periods which not only do not contain science but act on ideas entirely alien to it.

5. RELATIVISM, PRELIMINARY REMARKS

Assume that the inhabitants of such a culture lead a full and satisfying life, that they have ways of dealing with major problems, that some of the ways succeed while others fail but that the balance seems acceptable, both to the population itself and to an outsider who has not just studied their thought and their habits but has managed to incorporate both into his life. There are many cultures which satisfy this requirement. Classical scholars have explained how the Homeric heroes survived in a world full of strange and unfamiliar events. Their colleagues in the Latin section have made us familiar with the role of gods in a practical society such as ancient Rome. Travellers have marvelled at the amazing knowledge and the high moral standards of 'primitive' tribes and populations. Modern agents of development more than once had to adapt their own 'progressive' plans to local needs and ways of doing things which, they found, were better suited to the social and natural environment of the target population. We may even add groups and individuals who starve and castigate themselves - but with a clear purpose in mind. All these forms of life are bona fide candidates for Aristotle's principle - and relativists don´t hesitate to apply it. Atoms, they say, are not simply there - period; the Homeric gods are not simply fantasies - period. Atoms exist for the civilization that is built around them and sustains its inhabitants, while the Homeric gods exist for another civilization of comparable stability. Given what I have said so far, the inference seems inevitable. However, the situation is not quite that simple.

The reason is that the 'relativism' implicit in Aristotle's principle differs from the classical version. For a classical relativist things exist, are true, valid, good or evil - relative to a community that is united by common perceptions and standards. The mere existence of communities who look at the world in a certain way suffices for declaring that the opinions they call true are true relative to these communities. The 'relativism' of Aristotle's principle assumes in addition that life inside the reference communities has a certain quality - at any rate, people want to live this life and not another, they believe that this is how things are and they reject alternatives. Using Aristotle's principle we take such judgements at face value, i.e. we say that the preferred entities exist and have the properties ascribed to them. Realising that the principle applies not only to one case but to many, we treat the alternatives in the same manner. For example, we say that atoms and they alone exist, but we also say that the Homeric gods exist and that they are not bundles of atoms (an assumption made by Democritus and some atomists) and thus arrive at a contradiction. The epistemic relativist (as I shall call a philosopher who refers judgements to an individual, or a group, or an entire nation no matter of what quality) avoids the contradiction by inserting his 'for'; (s)he abandons Aristotle's principle. A person trying to retain its full force will concede that the entities populating the selected cultures and periods are infused with a reality that does not depend on a culture, or a point of view; trying to remain consistent (s)he will add that they are not intrinsic parts of that reality, but results of an interaction in which the reality responds to some of its more independent ingredients. The addition is relativistic: Being responds in more than one way and the response depends on the approach. It is objective: the approach alone does not decide the matter; some candidates (new mutations, incipient cultures, assemblies of matter about to form a galaxy) lack in robustness and soon dissolve. Is it also acceptable?

To start with, it is clear that the reality I introduced to free epistemic relativism from its lack of substance cannot itself become an object of knowledge. (It is the medium that connects all types of knowledge.) It shares this property with Kant's 'Ding an sich.' But like the 'Ding an sich' it also does a job: it gives robust cultures a property that is lacking in their more evanescent competitors. Radical empiricists reject entities which have no other function than to reconcile facts and philosophical prejudices. This is a possible move. I reply with a second application of Aristotle's principle: the 'prejudices' are deeply ingrained, they play as large a role in our lives as do more tangible phenomena and they are therefore candidates for a further, though not directly comprehensible form of reality (which, incidentally, can also be supported by a 'moral' argument - cf. item 4 of the next section).

Secondly, the point of view I have just sketched in no way conflicts with the so-called success of science. There is no uniform entity 'science' that might have this thing, success. What we have are different approaches, many of them in conflict with each other. Also, scientific success is a rare flower. Some approaches linger for years in a kind of limbo and then wither away, others are beaten down for centuries, are often declared defunct, but they survive and and eventually overtake their more successful rivals. The most superbly successful theories are not stable, they change which means that the success of their original form lasts only a short time. And even a longlasting success is infinitesimal, compared with the life span of homo sapiens, the earth, or the universe and thus might well be the result of a fluctuation coming from an entirely different medium.[7]

Thirdly, many scientific facts which are being discovered today arise as a result of sizeable interventions in the course of nature. The idea that they existed before the intervention and in exactly the form in which they appear is much less plausible than the idea that they are responses which would not have occurred without the intervention (this assumption also corresponds to a widely accepted interpretation of quantum mechanics).

It is important to realise that even the ontological relativism I just sketched may in the end turn out to be a chimera. The traffic between nations and the changes brought about by the intrusion first, of nomads and neighboring tribes or empires and then of Western Civilization make it very difficult to separate one form of life from another. Also Third World cultures - what is left of them - are very complex. Government officials almost everywhere wear a tie and a business suit (so, incidentally, do hired killers, except that their briefcases contain submachine guns); they welcome the investment of foreign capital and the technology, especially weapons technology, that goes with it. The airports of large cities, the hotels and the downtown shopping areas become more and more similar to each other. Does this mean that the ideology often connected with Western Civilization (secular humanism, for example) has now spread all over the world? It does not. Politicians and intellectuals keep their 'superstitions' and fall back on them in times of trouble (example: Haiti; in the West we have a multitude of Christian sects, Buddhism, remnants of Marxism - and so on). Besides, politicians and intellectuals are a very small minority. Populations in rural areas perpetuate older habits, partly out of inertia, partly by conscious effort even against tremendous odds. City slums with mixed populations grow in close proximity to high class neighborhoods, in New York and elsewhere. What is the 'culture' there? Is there a culture, or just a concentration of despair? Obviously, the matter cannot be decided from afar. All this means that even the ontological relativist's "for ..." in many cases either lacks a clearly circumscribed reference, or it is just an empty formula that can be attached to anything. But the same is true of scientific realism. Speaking of a single unified entity, 'science' which allegedly reveals a real and observer-independent world is nothing but a somewhat staid metaphysical romance. Ontological relativism may also be a romance but, I contend, a better one. I shall now present some of its consequences, in the form of a series of numbered assertions and comments.

6. ONTOLOGICAL RELATIVISM

1) There exists a world which underlies all things, events, processes. I shall call it the primal world or, simply, Being.
2) The primal world contains (produces) parts (individuals, groups, civilizations, species, galaxies etc.) which are relatively independent, and responds to them. Some parts fall back into Being (negative response) while others expand and stabilise (positive response). The problem of how a part of Being can become independent and yet remain a part of Being occurs in all forms of monism, from Parmenides to supergravity. It is occasionally defused by pointing out that Being, or God, is beyond knowledge and can only be hinted at by analogies, images, symbols. According to Pauli, quantum mechanics uses the same procedure. The wave function, says Pauli (op.cit., 15, footnote) is "a symbol that unites contradictory perceptual ideas" ("hat die Bedeutung eines die Gegensaetze der anschaulichen Vorstellung vereinigenden Symbols"); talk about Being is similarly symbolic, not informative; it unites different and, if generalised, contradictory manifest worlds.[8] The idea that Being might be incomprehensible is also suggested by the fact that, according to the orthodox interpretation of quantum mechanics, there is no such thing as the wave function of the universe (or even of a fly in front of me) and by the very different assumption that information about all there is, brains included, is bound to overload these brains. The idea that (uniform) Being produces (relatively) independent parts (emanations) occurs in Near Eastern Creation stories, in Hesiod, in neoplatonism, in neoplatonic versions of Christianity (Pseudo Dionysius Areopagita, for example) and in certain forms of pantheism. The idea that Being reacts to these productions (or anticipates them by special motions) occurs in the Lurianic Kabbalah (cf. Gershom Scholem, Sabbatai Sevi, Princeton 1973, 25ff) and also in Christianity: God sent his son to atone for the crimes of one of his creations. Christian philosophers had great difficulties to reconcile the relative independence of, say, humans, or of the laws of nature with divine power and divine will (Duns Scotus and Ockham against St. Thomas - but cf. e.g. St. Thomas, Summa Theologica, question viii, first article: -'Whether God is in all things' and the following three articles). Scientists face exactly the same problem. For if there is an objective reality whose nature is independent of individual opinions, historical traditions and even of the existence of human beings, then how can these peripheral creatures who are hampered by so many idiosyncracies ever hope to get a glimpse of it? The more objective reality becomes, the more urgently we need some form of revelation to understand it.

3) Not all responses of Being are stable (in human terms) and can support an (again in human terms) at least partly satisfactory life (not all relatively independent parts of Being are robust and likely to grow). Matter accumulates and disperses, biological mutations appear, linger for a while and are never seen again, cultures lacking in strength are absorbed by larger units and so on. Responses (to human action) which retain their independence are perceived and described as worlds put together from special entities (material objects, animals, divine beings). I shall call such worlds manifest worlds. Manifest worlds are the cultural analogue of ecological niches. Being overwhelmed with what immediately affects them the inhabitants of a particular manifest world often identify it with Being. Scientists, for example, think that the basic laws of physics characterise Being itself, not a special province of it. They assume that the products of their highly idiosyncratic activities existed before and independently of them and in precisely the form outlined in the most recent textbooks. Speculation along the lines described above undermines this provincialism and opens a wider, though not at all well-defined perspective.

4) Since manifest worlds are part of Being and can be stabilised, improved, removed by suitable actions (though unpredictably so - see below) it is possible to have absolute convictions and to make absolute decisions. There is no guarantee that they will succeed (Being is not like the Mosaic God who suggested a contract and felt bound by its conditions), but they at least make sense because they are no longer castrated by the relativistic "for ...." This removes a difficulty noticed by Marcello Cini and Bas van Braassen. (We can also start from the postulate of the validity of absolute convictions and infer a primal world; Kant proceeded in a similar way.)

5) As already stated, it is not possible to draw a distinction between Being and its manifest parts. Drawing such a distinction would attribute concrete properties to Being. Even the most profound distinctions are features of particular manifest worlds, not of Being. Commenting on a similar property of Jungian psychology Wolfgang Pauli wrote (op. cit., 115): "... since every expansion of consciousness ... must have a back effect on the unconscious, we have to expect that there will be a 'problem of observation' with regard to the unconscious which is analogous to the problem in atomic physics, but contains far greater difficulties. These difficulties are bound to show themselves in logical paradoxes when one attempts to give a conceptual account of the unconscious." Replace 'conscious' by 'manifest' and 'unconscious' by 'primal'- and the situation is exactly as described.

6) 3 and 5 imply that Being cannot be known. Assertions apparently about Being are ways of presenting manifest results. The religions and

philosophies I mentioned towards the end of 1., try to clarify the situation and the role of language it implies. Having researched a particular manifest world which is spread out in time, we can give a historical account of the results ("in 1990 supergravity seemed like a good thing"), we can try to stabilise them, but we cannot regard them as parts of Being. There is no theory of Being, there is not even a history of the ways in which the manifest worlds emerge from Being, there are only the histories of these manifest worlds themselves embedded in wider manifest worlds whose intellectuals may try to explain their historical, social, theological role or show to what extent they are 'approximations' (cf. Pauli's discussion of the wave function in 2).

7) If such wider worlds have not yet arisen then the transition from one manifest world to another cannot be accounted for except by agencies (creativity, divine intervention, absolute accident, removal of prejudice) which give color to our efforts without explaining them. Statements such as "The Great Pan is Dead" (Plutarch), or "God is dead" (Nietzsche) describe the decay of particular manifest worlds - and here the term "death", though somewhat melodramatic is indeed appropriate - they do not touch Being which lies beyond events.

8) Laws of nature (which are structures of particular manifest worlds) share with social laws their dependence on historical conditions, not in the sense that they are discovered because of these conditions but exist independently of them, but in the very different sense that they and the social structures together form a particular manifest world.

9) Conversely, the ideas preparing, articulating and exploiting this world, being part of primal reality, also have an (unknown) inhuman component.

10) Since the events that have been discovered by science, i.e. the big-bang universe, its ingredients, laws, developments and other manifest facts are the response of an (unknown) material to human action we may say, in the same symbolic manner that applies to all statements about Being, that Nature is a work of art, formed from the substance of Being and subjected to its whims. The difference between such an artwork and pictures, poems, symphonies is (a) that generations worked it and (b) that the material, Being, remains unknown to the artists. They need grace to succeed and revelation to understand what exactly it is they have achieved.

11) While the terms of the sciences and of commonsense are adapted to a particular manifest world but incapable of grasping its relation to Being, or of even hinting at the problem, religious thinkers, poets, mystics have tried to develop ways of talking that lead beyond such boundaries. There exists now a physical theory, the quantum theory which encouraged similar efforts (cf. the quotations in 2 and the Jung-

Pauli correspondence). Combining both trends, we may arrive at a world view which, though not rational or cognitively meaningful in the sense of neopositivism, yet gives a better account of the human condition than purely scientific constructions, especially when the latter are still dominated by the spirit of 19th century science. Taken by themselves, the results of the so-called 'Sciences of Man' are as incomplete as landscape painting.

7. THEORIES AS SYMBOLS AND AS HISTORY

I started the essay by denying that there could be a theory of humans, society and even of nature. I now want to clarify this statement. There are of course sentences and formulae which are called theories, but they function in a rather unusual way. They are either tendencies of a particular manifest world which are as evanescent as this manifest world itself, or they give a symbolic account of how the manifest world might be related to Being. In the first case they are really histories, but the historical element is contained in their use, not in their formulation (which remains stable and thus creates the illusion of an objective stability). To clarify the situation, let us assume that we want to formulate a theory of reason. Now reason is an object (for example, it is an object of study) and it is an agent. The nature of the object 'reason' becomes clear after the agent 'reason' has acted. A theory of reason, when stabilised, restricts the possible actions of reason - it identifies reason with the mirror image of its latest actions. What if reason does not conform to the restriction? Then, the theorists say, the theory must be adjusted. But the adjustment occurs at every turn of history which means we have a theory only in words; what we actually have is a development, a story. This is overlooked by philosophically inclined scientists like Bohm, Prigogine, Jantsch, Maturana, Varela, Thom who claim to have found universal principles that can accommodate wars, artistic revolutions, the disappearance of species, marriage, divorce, creativity and many other things. For how is the miracle achieved? We have a narrow theory that describes some special phenomena in a surprising way, i.e. we start from a solid scientific result. Next, the formulation of the result is generalised by adding some loose parameters: the enterprise is not narrow-minded, but 'philosophical'. But the parameters receive definite values only after the events they are supposed to describe have occurred which means that again we don't have a theory, only a rather cumbersome and quite misleading way of telling a story. For 'reality', or 'nature', the argument is exactly the same. Reality is an object research, and it is an agent - all scientists are subjected to it. Neither the agent nor the researcher is bound by the ideas which both produce at a certain stage. If they happen to be so bound, then this is a historical accident duly recorded by the historical dimension of the statement misguided scientists and philosophers are interpreting as a universal law of Being.[9]

Prof. Dr. Paul Feyerabend
Institut für Philosophie/Abt. XII
ETH Zürich
8092 Zürich, Suisse

NOTES

1 Diels-Kranz (eds.), *Fragmente der Vorsokratiker*, 6th edition, Berlin 1951, B7, 7.

2 Aristotle used this result to develop a fascinating theory of continuity. The theory implies, among other things, that moving objects have no well defined length in the direction of their movement - cf. my book *Farewell to Reason*, chapter 8, proposition 14.

3 Einstein to Besso, quoted from P. Speziali (ed.), *Albert Einstein, Correspondance avec Michele Besso*, Paris 1979, p. 312; cf. also p. 292. The preceding quotation is taken from H. Weyl, *Philosophy of Mathematics and Natural Science*, University Press, Princeton 1949, p. 116. For irreversibility cf. Stephen G. Brush, *The Kind of Motion we Call Heat*, Vol. 2, North Holland Publishing Company 1976, esp. p. 650. For the attitude of molecular biologists cf. P. Fischer, *Licht und Leben*, Universitaetsverlag Konstanz 1985, p. 229. I. Prigogine explains the situation in *From Being to Becoming*, W.H. Freeman & Co, San Francisco 1980, while I. Prigogine and I. Stengers, *La Nouvelle Alliance*, Paris: Gallimard 1979 contains historical material on the persistence of Parmenidean tendencies.

4 Diels-Kranz, op. cit., fragment B1.

5 It started looking like an argument - but it was not an argument, i.e. it was not an objective invitation to drop accepted opinions. Philosophers who believe that the noise they make can change the world quite naturally praise Xenophanes for his 'destructive criticism' (W.K.C. Guthrie, *A History of Greek Philosophy*, Vol. i, Cambridge 1962, p. 370), or his "acute criticism" (M. Eliade, *Histoire des croyances at des idees religieuses*, Vol. ii, Paris 1978, p. 497) or his "discovery that the Greek stories about the gods cannot be taken seriously because they represent the gods as human beings" (K.R. Popper, *Auf der Suche nach einer besseren Welt*, Munich 1984, p. 218). Older philosophers such as Nietzsche and Hegel said the same. They overlook the fact that thought, taken by itself, lacks both content and power unless acting in harmony with extra-intellectual agencies.

6 Quoted from K.V. Laurikainen, *Beyond the Atom*, Springer, Berlin-Heidelberg-New York 1988, p. 60. For the whole problem cf. the discussion at the symposium "The Copenhagen Interpretation 60 Years after the Como Lecture", 6-8 August 1987, Poensuu, Finland., report published by the department of Physical Sciences, University of Turku, SF-20500 Turku 50, Finland, March 1988, especially the second section. Cf. also chapter 16 of W. Pauli, *Physik und Erkenntnistheorie*, Braunschweig 1984 (attempts to unite science with steps toward salvation), p. 15 of the same book (argument against a "losgeloester Beobachter") as well as the most revealing Pauli-Jung correspondence about to be published by Springer.

7 Cf. Boltzmann, *Vorlesungen ueber Gastheorie*, Vol. ii, Leipzig 1912, section 90.

8 Heisenberg speaks in a somewhat different way (*Der Teil und das Ganze*, Piper, Munich 1969, pp. 284ff). Quantum theory, he says, contains very precise abstract laws. These laws "give us a uniform description of atomic phenomena which, however, can look very different when an application to experiments leads to a translation into the natural language. Quantum theory is a wonderful example of the fact that though one may be completely clear about a state of affairs, one may yet know that one can only talk about it in images and similes". The relation between faith and rational discourse in religion cannot be described more clearly.

9 A similar conclusion can be reached on the basis of more recent investigations in the history of science. Cf. my article 'Realism and the Historicity of Knowledge' in the *Journal of*

Philosophy of 1989 as well as my contribution to *Festschrift fuer Adolf Gruenbaum*, 'The End of Epistemology?', forthcoming.

ETHICS IN SCIENCE - SUBSTANCE OR RHETORIC?

JÜRGEN MITTELSTRASS

Ethics is a hot commodity - with the general public, with philosophy and with the sciences. The reason for this lies less in the increased performance of ethical reflection than in a growing lack of orientation that has embraced even the sciences, or at least our dealings with scientific results. The great interest which greeted Hans Jonas with his "Attempt at an Ethics for Technological Civilizations" (the subtitle of his book Das Prinzip Verantwortung, Frankfurt: Suhrkamp, 1979) bears eloquent witness to the fact that the modern world needs ethics - and apparently has not yet found the right ethics.

How much ethics and which ethics does the modern world need? How much ethics and which ethics does science need - natural science just as much as the science of man and society? In the following remarks I shall deal especially with this second question (with somewhat provocative intent), and I will concentrate in an exemplary manner in particular on the relation of ethics and medicine. For in medicine all the sciences of man are connected in a very concrete manner for the future of man and society.

1. SCIENCE AND PANDORA'S BOX

The modern world is, in its structures and forms of life, the expression of the scientific and technical mind. Science - and technology, too - is everywhere today. Wherever we go in the modern world, we always find that science and technology were already there. Whatever we know, the scientific mind knows it better; whatever we do, the technical mind guides our hand. Without science and technology the world today would be uninhabitable. Science and technology, thus, are not something that we could dispense with or exchange for another, non-scientific and non-technical form of life. Those who think this is possible not only deceive themselves, they also envision a situation in which the world and mankind have lost their rational essence.

Nonetheless, science in particular has today become somewhat notorious. The ivory tower in which it resided according to traditional notions and from which it blessed the world with discoveries and insights, has lost its luster. We hope that science can lead us back to a paradise lost (naturally much better furnished than the biblical one), but we also mistrust it, ever since Mephistopheles took possession of Faust. The reason for this mistrust is however by no means simply literary or academic, but rather quite concrete. It is that the progress in which science manifests itself has no immanent standard or measure. It goes where it will. It changes our world, it orients our world, but it also leads to problems of orientation. Furthermore,

J. Götschl (ed.), Revolutionary Changes in Understanding Man and Society, 269–277.

scientific developments and technical developments are mutually dependent. The progress of one drives the progress of the other forward. A standard for this development is not in sight. Should there be an inner standard of science and technology, then it would be to overstep any measure. A standard or measure is always a limit or limitation. But scientific and technical rationalities define themselves precisely by the temporary nature of every limit.

It is, therefore, not the performance or the capacity to perform of scientific and technological rationalities that is called into question, but rather the "inner reasonableness" of these rationalities, that is, a standard that evaluates not only in terms of scientific and technological dynamics but also in terms of compatibility with and support of well-grounded goals and purposes. Rationality alone, precisely in its scientific and technical form, is not a sufficient condition for a humane world. The philosophical tradition has tried to give expression to this through the distinction between *reason* and *understanding*. Understanding provides the rationality that changes the world; reason provides the normativity that guides and judges the changes according to a reasonable standard.

Realisation of this kind of standard is what is problematic today. The ways of science in research and application become harder to oversee and supervise and also become more dangerous. One indication of this is the increasing and increasingly problematical entanglement of science and law, for instance in biology and medicine. Science's intemperance with its own measure affects its results as well. We have already reached the point at which the effort needed to deal with the consequences of scientific research is no smaller than that invested in the research and in drawing the consequences. Are the consequences getting out of control? Among the consequences of scientific and technical rationality that make the world not only more hospitable but also make life in this world harder to survey and more full of problems are the following:

1) In *progress*, especially in so-called technological change closely connected to scientific developments, there is no longer a recognizable subject or agent. Technological change, which belongs to the dynamic of the modern world, has to a large extent become an anonymous affair. Progress has long since become incomprehensible to the individual, it transpires behind his back. This is why it also cannot simply be stopped and why responsibilities cannot be easily and definitely ascribed. This applies in the area of ecology just as much as in that of medicine and in all other areas in which the modern world displays itself. Responsibility thus becomes responsibility in regard to developments for which in a strict sense no one is responsible. The consequence is that structures of

responsibility in general are on the decline, so too, in scientific contexts.

2) Knowledge dissolves in information. The modern world dresses itself up in communicative structures as a world of information, modern society has become an information society. This, too, creates problems. Information, though based on knowledge, increasingly puts itself in the place of knowledge. Modern man is less and less able to comprehend what is made available to him as information, to distinguish what in the medium of information is knowledge and what is opinion. The consequence is a loss of epistemic independence. And this means, to the extent that information grows and the knowledge that one has acquired and mastered oneself decreases, to this extent does the incomprehensibility of the scientific-technological relations in which modern man lives grow. We all become giants in information and dwarfs in knowledge.

3) Modern societies are transformed into *expertocracies*. The expert becomes the symbol of a world increasingly dependent on the employment of science and technology. Only the expert seems able to connect knowledge to practice, both social and individual. But this fosters the illusion that all problems can always be solved "technically", that is, that they are the kinds of problems that fall within the jurisdiction of experts. And this is, once again, a misunderstanding. It arises when man considers himself exclusively through the mirror of a world he has created with his scientific and technological rationalities. This world however does not explain man.

4) Alongside the *appropriation of nature*, which brings its own (environmental) problems, we find the *appropriation of man's own nature*. With the rapid development of biological research (for instance, genetic engineering) man has come into a position to change himself, that is, to do to his own nature what he has long since been doing to "external" nature. What was once considered inaccessible, is being resolved into elements that are technologically at least partially accessible through the interventions into the genetic identity of humans. Herein lies not only a completely new quality of scientific and technical progress but also an immediate threat to mankind. People can, then, be manipulated not only ideologically but also biologically. The subject and agent of progress becomes its object.

What I have described in four examples here as consequences of scientific and technological rationalities can also be expressed as the tendency of these rationalities to take on a life of their own and of mankind in this development to confront itself ever more powerfully. The world, as a scientific-technical world, a work and artefact of man, itself takes on productive characters. It works upon man, it changes his environment and it changes him. But not always for the best of mankind and its world. Even developments such as those in the areas of energy and transportation, with which the concept of progress is bound, contain ambivalent elements. The danger to the biosphere through the gradual transformation of the earth into a garbage dump and the unmistakable burdens of toxic wastes may serve as examples. In Greek antiquity Pandora's box followed Prometheus' fire. Out of Pandora's box spring all evils; only hope remains inside. Will the Greek myths be confirmed in the modern world? Unmistakably, those who believe this world is lost and hold science in particular responsible for its fate are gaining an audience and (social) influence. The Zeitgeist paints in dark colors and science is a favorite subject.

Now, I too don't want to deny that science not only solves problems but also creates them. The danger to the biosphere is among others the effect of successful scientific (and technological) rationalities (for instance fluorocarbons). However, it would be naive to believe or attempt to persuade others that an injunction on science and technology limiting their power and efficacy would suffice to direct the world into calmer waters without at the same time making it lose its rational essence. It is true that a dilemma exists today between scientific-technological progress and its consequences, between technological change and technology's impact; but it is false that this dilemma can be resolved by scientific and technological asceticism. It would be straightforwardly mistaken to think that the problems of our world, including those due to its scientific essence, could better be mastered with less science and less technology. In fact, the opposite is the case: A marked cutback of scientific research and technological development would soon make the modern world incapable of acting or reacting. The problems of this world will not stand still if we stand still; and - whether this suits one's *Weltanschauung* or not - these problems won't be avoided by less science and technology, but only by more science and technology - or at least made more bearable.

With this we have formulated *a scientific imperative*, that is, a research commandment. While obeying this commandment will not deliver us once and for all from the dilemma of research and its (unwanted) consequences, its ineffectiveness would allow our world to sink ever deeper into self-caused and other problems. Thus, hunger in a world that has no control over its population growth, on an earth whose natural carrying capacity is long since exhausted, will not be conquered by political righteousness or ideological fundamentalism, but only - at least for the time being, that is, until other means, such as enlightenment and a rational world politics have a chance to act - with more science and technology. The alternative to a prudent further development of the modern world would be pure cynicism in light of the consequences of the renunciation of such a further development, which would likely result in the death of hundreds of millions. This means: in view of the self-caused problems alone, among them toxic wastes and overpopulation, there is no real alternative to the modern world. And therefore, there is also no alternative to research and development.

2. ETHICS OF SCIENCE?

In this context - of a prudent further development of the modern world and a formulation of a scientific imperative - we often hear ever more pronouncedly the keyword "ethics", not only in political discussions but also in scientific discussions. A new (or old) ethics is supposed to accomplish what science itself no longer accomplishes, what it cannot accomplish at all based on the rationality that defines it. Science, itself without internal measure, leads to ethical (and other) problems, it doesn't solve them. One example of this is the interface of genetic engineering and reproductive medicine when the former is applied in human genetics. Here, as in the earlier examples (that is, anonymity of progress, information society, rule by experts, appropriation of nature including our own), scientific and technical rationalities run away from life-world orientations and their ethics. Is a new ethics of science the answer to our problems? Many today think it is. My answer, however, is: No.

We should note first that man, who takes his world in his own hands, who makes it his own work, that is, the inhabitant of the modern world, does not have fewer ethical problems than his ancestors, who had to arrange themselves with a much more natural world, but rather more. It is not only that our life-world problems do not in fact disappear when our scientific problems are solved, but that it is precisely the solutions to the scientific problems that cause new ethical ones. Such a problem, in the sense of the above mentioned appropriation of our own nature, occurs whenever the distance between man as researcher and man as object studied becomes too small and research begins to turn upon man. Informational self-determination, reproductive medicine, genetic engineering are keywords for this diminishing distance. At the same time, it is not at all clear how long ethical boundaries that have held up till now will withstand scientific and technological developments; boundaries, for instance, between prenatal diagnosis and its application as selectional diagnosis, between the sequencing of the human genome and its possible application, between genetic therapy and germ line therapy. In all these cases research creates problems that themselves have no scientific solutions, that cannot have purely scientific solutions. Ethics becomes the ever-longer shadow cast by science and technology in our world.

If science raises ethical problems that it cannot solve itself and (at least with regard to its consequences) is subject to evaluations that it cannot supervise itself; if, in other words, scientific (and technical) rationality does not itself say what science (and technology) may or may not do, then apparently a standard of scientific (and technical) progress can only be a standard imposed from outside. Can a new (or old) ethics of science provide this standard?

On this point the hopes at present placed in such ethics greatly exceed the possibilities of realization. It is possible that the call for an ethics of science even leads us in the wrong direction. The reason for the negative answer indicated earlier is the following: science constitutes a special form of knowledge but not a special form of sociality with a corresponding ethics. Those who think this way not only confound the methodologically grounded rationalities of knowledge with ethically grounded social rationalities, but they are also guided by a false notion of ethics. I am referring to the notion that for all problems and all situations there is an ethical answer - indeed, *the* ethical answer - in other words: the notion that ethics is a discipline that in all imaginable cases puts us in a position to say what is morally justified and what is not. Such a notion is (some might say, unfortunately) misleading. It confuses ethics with a textbook and the solution of ethical problems with the application of theorems.

Ethics does not consist in the possession of such absolutes, and there can therefore not be an ethics of science as distinguished from other ethics. Furthermore, the scientist is the subject of moral claims not *qua* scientist but *qua* citizen, so that ethics is always a *citizen's ethics*. Ethics cannot be divided socially into a scientific ethics as special ethics for scientists, on the one side, and a non-scientific ethics as standard ethics of society, on the other. Strictly speaking, there are no closed ethical worlds in which respectively only a special ethics applies.

This caution is directed not only against exaggerated hopes for a proper ethics of science but also against the notion that the scientist is responsible for more than the citizen. The scientist does indeed have a special responsibility insofar as the scientific mind cannot in principle be supervised by the non-scientific mind and insofar as modern society is dependent on the capabilities of the scientific mind, but this responsibility does not manifest itself in a special ethics. What is necessary, instead, is a particular ethos such as that long since practiced with social effect in the professional ethos of the physician. All rules, all norms that might be promulgated for the practice of science to strengthen its responsibility and secure its rationality would be a waste of time if such an ethos did not already exist. The freedom of science (in research and teaching) as a constitutionally protected value and in regard to the problem of responsibility just discussed has as its prerequisite that this scientific ethos is, in general, not corrupt. And little more can be said at this point about an ethics of science.

3. ETHICS AND MEDICINE

For a long time inquiries into the ethics of science were directed primarily at physics and biology. It was overlooked that not only the *theoretical* sciences but also the *practical* sciences immediately related to action, like medicine, are at issue here. The fact that this is not generally clear is in turn due in the case of medicine especially to the notion that medicine is an applied natural science and thus itself a theoretical science. But even the very notion is misleading. If it were right, it would mean that the ethical problems of medicine would have to be decided outside of medicine.

In connection with the question of a science ethics, what is much more important than the old question whether medicine is a science, is the structural transformation that has occurred in medicine. It is based on the increasing permeation of medical practice by science and technology - something that will not exactly be news to a radiologist. In this context, medicine has acquired dispositional power over man to an extent that was once unimaginable, and hence, what I have said about the accessibility of the nature of man in biological research also applies to medicine. Thus, in medicine, much more concretely than in other disciplines, even science-related actions unavoidably acquire ethical dimensions. This is even the rule here, not just a borderline case. At the same time, the case of medicine makes it clear that the consequence of the increasing permeation of all areas of human life by science is not that ethical standards are dispensable but rather that they are unavoidable. The more knowledge and know-how we possess, the more urgent are orientations that cannot be provided either by knowledge or know-how. And, furthermore, a practice being transformed by science and technology will be ever less able to develop and instill such orientations in the form of practical knowledge.

A number of developments in which the researcher in medicine becomes a potential culprit, make it clear that we are dealing not just with "abstract" things of interest only to the philosopher of science. Examples of this are the disregard for patients' rights in pursuit of research, such as happened with the injection of virulent cancer cells presented as a therapy-oriented immunization test in the Chronic Disease Hospital in New York; or human embryo research, when this crosses the boundary to experiments with human life as in the case of the production of embryos for research purposes; or the already mentioned prenatal diagnosis, when diagnosis is used for selection. In all cases of this kind, but also, for example, in the case of cancer therapy where radiology often causes serious damage and where there is no technical answer to the question of which is worse, the effects of the disease or the effects of the therapy, scientific research has immediate ethical or legal implications. These arise not only because scientific knowledge in the form of basic research is applied in a problematical manner but also because, for instance in genetic research, basic research and applied research are meshed together so that they are increasingly indistinguishable. For this, too, medicine is a prime example, since it not only applies the results of the biology of the gene but, in human genetics, also engages in it.

This view of things is, as a matter of fact, not restricted to the perspective of research but also holds for the perspective of the patient. The more intensive medical technology becomes, the more intensively must the physician as a person deal with the patient. Wherever the physician disappears behind the apparatus, sickness, suffering, and healing lose their human measure and standard, the machines transfer their structures to the sick human being, who learns to consider himself even when healthy only as a well running machine. So far, medicine also retains more prominently than other scientific disciplines a remembrance of the fact that the world cannot be arbitrarily permeated with science and that the real measure of science is an *ethical* measure. And once again we see that, even in the case of medicine, there is in a strict sense no special ethics but only concretizations of a general ethics, namely of the citizen's ethics, for medically relevant situations. Besides this, there is also a canon of rules for medical actions that determines not so much an ethics as an ethos in the sense explained.

4. THE FUTURE OF PRACTICAL REASON

If the remarks presented here on taking an ethical measure of science go to the heart of the matter and don't just hit the usual rhetoric, then the real problem lies not in the question whether an ethics of science is possible or not, but rather in the fact that an ethos which is indispensable in science, too, is no longer self-evidently anchored in a non-controversial universal ethics, namely the general citizen's ethics. In ethical questions as well, we no longer live in a closed world.

The dynamics of change, whose engine in the modern world is science, extends also to man and his orientational structures, including ethics. These orientations have long since ceased to be anything natural. A creature that makes its world with the help of science and technology and constantly subjects this world to changes cannot hope for or count on the naturalness of its orientations. These orientations, too, especially insofar as they are ethical, are not simply there, least of all there as part of the natural endowment of humans. They are rather "made" and they change to the extent that the modern world, the man-made world changes. In other words, to the extent that the scope of action of humans is widened by science and technology, the forms of orientation, including the ethical forms, are also subject to developments.

Man will never get into a situation in which science finally explains everything and at the same time also creates those orientations according to which man should and can live. And this is less a disappointment than a glimmer of hope. It says that man, without losing his modernity, can avoid his own transformation by science and thus, too, his appropriation by the world that he has made. This, then, is the real challenge to modern man presented by his world. The way to meet this challenge, however, does have something to do with ethics - with ethics not only as an authority that regulates the relations of humans among themselves and gives scientific and technological developments boundaries and measures, but also with ethics as a medium in which man and his world can once again find a unity. For this we need - according to a genuine and very concrete insight of philosophy - not theoretical but practical reason. Reason, too, not conceived as something "already there", already finished as a philosophical system or social reality. Reason, rather, as permanent enlightenment and as the declared will to live once again in a humane world.

Science and technology will be a part of this world. To imagine such a world without them would be naive and would mean wanting to return to mythic forms of life. It will, however, be science and technology within the bounds of practical reason, not in the overtaxing role of this reason itself. Idylls and science fiction lie close to each other when the measure of the world is sought not in the practical reason of man but in promises either of the simple life or of pure knowledge. This search is to be discouraged, even though reason, the production of a human world, is an infinite task. The ever renewed attempt to accomplish this task is the real *conditio humana*. And on this, I hope, even the scientist and the philosopher will be able to agree with all the others who today still adhere to the program of the Enlightenment, the "project of modernity".

Prof. Dr. Jürgen Mittelstrass
Philosophische Fakultät
Universität Konstanz
Universitätsstraße 10, 78434 Konstanz, Germany

GPSR Compliance

The European Union's (EU) General Product Safety Regulation (GPSR) is a set of rules that requires consumer products to be safe and our obligations to ensure this.

If you have any concerns about our products, you can contact us on ProductSafety@springernature.com

In case Publisher is established outside the EU, the EU authorized representative is:

Springer Nature Customer Service Center GmbH
Europaplatz 3
69115 Heidelberg, Germany

Batch number: 10824484

Printed by Printforce, the Netherlands